Extending Performance of Concrete Structures

Proceedings of the International Seminar
held at the University of Dundee, Scotland, UK
on 7 September 1999

Edited by

Ravindra K. Dhir
Director, Concrete Technology Unit
University of Dundee

and

Paul A. J. Tittle
Research/Teaching Assistant, Concrete Technology Unit
University of Dundee

ThomasTelford

Published by Thomas Telford Publishing, Thomas Telford Limited, 1 Heron Quay,
London E14 4JD.

URL: http://www.t-telford.co.uk

Distributors for Thomas Telford books are
USA: ASCE Press, 1801 Alexander Bell Drive, Reston, VA 20191-4400, USA
Japan: Maruzen Co. Ltd, Book Department, 3–10 Nihonbashi 2-chome, Chuo-ku, Tokyo 103
Australia: DA Books and Journals, 648 Whitehorse Road, Mitcham 3132, Victoria

First published 1999

The full list of titles from the 1999 International Congress 'Creating with Concrete' and
available from Thomas Telford is as follows

- *Creating with concrete*
- *Radical design and concrete practices*
- *Role of interfaces in concrete*
- *Controlling concrete degradation*
- *Extending performance of concrete structures*
- *Exploiting wastes in concrete*
- *Modern concrete materials: binders, additions and admixtures*
- *Utilizing ready-mixed concrete and mortar*
- *Innovation in concrete structures: design and construction*
- *Specialist techniques and materials in concrete construction*
- *Concrete durability and repair technology*

A catalogue record for this book is available from the British Library

ISBN: 0 7277 2820 2

Printed and bound in Great Britain by MPG Books, Bodmin, Cornwall

PREFACE

Concrete is the key material for Mankind to create the built environment, the requirements for which are both demanding in terms of technical performance and economy and yet greatly varied from architectural masterpieces to the simplest of utilities. This presents the greatest challenge and the question is how best to advance concrete and create imaginatively.

In response, the Concrete Technology Unit (CTU) of the University of Dundee organised this Congress following on from its established series of events, namely, Concrete in the Service of Mankind in 1996, Concrete 2000: Economic and Durable Concrete Construction Through Excellence in 1993 and Protection of Concrete in 1990.

Under the theme of Creating with Concrete, the Congress consisted of five Seminars: (i) Radical Design and Concrete Practices, (ii) Role of Interfaces in Concrete, (iii) Controlling Concrete Degradation, (iv) Extending Performance of Concrete Structures and (v) Exploiting Wastes in Concrete, and five Conferences: (i) Modern Concrete Materials: Binders, Additions and Admixtures, (ii) Utilising Ready-Mixed Concrete and Mortar, (iii) Innovation in Concrete Structures: Design and Construction, (iv) Specialist Techniques and Materials for Concrete and Construction and (v) Concrete Durability and Repair Technology. In all, a total of 421 papers were presented from 67 countries.

The Opening Addresses were given by Mr Henry McLeish, MP, MSP, Minister for Enterprise and Lifelong Learning, Scotland, Dr Ian Graham-Bryce, Principal and Vice Chancellor of Dundee University, Mrs Helen Wright, Lord Provost, City of Dundee and Professor Peter Hewlett, President of the Concrete Society. This was followed by four Opening Papers by leading international experts; Dr Bryant Mather, US Army Corps of Engineers, Professor Charles F Hendriks, Delft University of Technology, Netherlands, Dr Bjørn Jensen, Danish Technological Institute, Dr Oliver Kornadt, Philipp Holzmann AG, Germany, Professor Jurek Tolloczko, Concrete Society, UK, Mr Michael Téménidès, CIMBÉTON, France and Professor Yves Malier, Ecole Normale Superieure de Cachan, France. The Closing Address was given by Professor John Morris, University of the Witwatersrand, South Africa.

The support of 20 International Professional Institutions and 31 sponsors was a major contribution to the success of the Congress. An extensive Trade Fair, participated in by 50 organisations, formed an integral part of the Congress. The work of the Congress was an immense undertaking and all of those involved are gratefully acknowledged, in particular, the members of the Organising Committee for managing the event from start to finish; members of the International Advisory and National Technical Committees for advising on the selection and reviewing of papers; the Authors and the Chairmen of Technical Sessions for their invaluable contributions to the proceedings.

All of the proceedings have been prepared directly from the camera-ready manuscripts submitted by the authors and editing has been restricted to minor changes where it was considered absolutely necessary.

Dundee
September 1999

Ravindra K Dhir
Chairman, Congress Organising Committee

INTRODUCTION

In addition to possessing adequate mechanical properties to ensure the serviceability of structural members, concrete must be durable, i.e. it must resist physical and chemical attack and protect embedded steel reinforcement from corrosion throughout its required service life.

Historically, the durability of concrete structures has not received sufficient consideration, and too much reliance has been made on indirect factors such as grade and water/cement ratio. As a result, the repair and strengthening of concrete elements has become an important part of the concrete construction industry. Many techniques widely used today are intrusive and costly, and there is the need to develop technologies that are more appropriate. One of the most promising methods involves addition of externally bonded steel plates, with the parallel development of high performance adhesives and anchor technology. More recently, fibre reinforced plastic plates have been used, which offer a number of advantages over traditional steel plates, although they can be overly costly for many applications.

Whilst current specifications for concrete, e.g. pr EN 206, are generally adequate to ensure the durability of the concrete itself, corrosion of steel reinforcement in structural members remains a major problem. Innovative solutions are increasing sought, including increasing the resistance of concrete to penetration of harmful substances, controlling tensile cracking, and using non-corrodable reinforcements.

Analysis and testing of real concrete structures is required to ensure reliable transfer of systems developed in the laboratory, and can provide early warning of durability problems. Practical applications for improving concrete durability are also being developed, such as controlled permeability formwork liners.

The Proceedings of this Seminar, *'Extending Performance of Concrete Structures'*, dealt with this subject area under two themes: (i) Materials Development and (ii) Practical Applications, with the aim of allowing in-depth focussed discussions on both topics. The Seminar started with a Leader Paper and both themes opened with a Keynote paper, presented by the foremost exponents in their respective fields. There were 27 papers presented during this International Seminar, which are compiled into these Proceedings.

Dundee   Ravindra K Dhir
September 1999   Paul A J Tittle

ORGANISING COMMITTEE
Concrete Technology Unit

Professor R K Dhir, OBE (Chairman)

Dr M R Jones (Secretary)

Mr M D Newlands (Joint Secretary)

Professor P C Hewlett
British Board of Agrément

Dr N A Henderson
Mott MacDonald Ltd

Professor V K Rigopoulou
National Technical University of Athens, Greece

Dr S Y N Chan
Hong Kong Polytechnic University

Dr N Y Ho
L & M Structural Systems, Singapore

Dr M J McCarthy

Dr M C Limbachiya

Dr T D Dyer

Dr K A Paine

Dr T G Jappy

Mr P A J Tittle

Mr J C Knights

Mr S R Scott (Unit Assistant)

Miss A M Duncan (Unit Secretary)

v

INTERNATIONAL ADVISORY COMMITTEE

Dr H M Z-Al-Abideen
Deputy Minister
Ministry of Public Works and Housing, Saudi Arabia

Professor M S Akman
Emeritus Professor of Civil Engineering
Istanbul Technical University, Turkey

Dr R Amtsbüchler
Manager-Technical Services
Blue Circle Ltd (South Africa), South Africa

Professor C Andradé
Director
Institute of Construction Sciences, Spain

Professor J M J M Bijen
Director
INTRON B.V., The Netherlands

Professor A M Brandt
Head of Section
Polish Academy of Sciences, Poland

Dr J-M Chandelle
Managing Director
CEMBUREAU, Belgium

Professor P Helene
Head of Civil Construction Engineering Department
University of Sao Paulo, Brazil

Dr G C Hoff
Senior Engineering Consultant
Mobil Technology Company, USA

Professor I Holand
Senior Research Engineer
SINTEF, Norway

Professor B C Jensen
Director
Carl Bro, Denmark

Professor S Mirza
Professor of Civil Engineering and Applied Mechanics
McGill University, Canada

Professor S Nagataki
Professor of Civil Engineering and Architecture
Niigata University, Japan

Professor H Okamura
Vice President
Kochi University of Technology, Japan

Professor E A e Oliveira
Director
Laboratório Nacional de Eng Civil, Portugal

Professor J-P Ollivier
Director of LMDC-INSA
LMDC, France

INTERNATIONAL ADVISORY COMMITTEE
(CONTINUED)

Professor R Park
Professor of Civil Engineering
University of Canterbury, New Zealand

Mr S A Reddi
Managing Director
Gammon India Limited, India

Professor H-W Reinhardt
Head of Construction Materials Institute
University of Stuttgart, Germany

Professor R Rivera-Villarreal
Chief of Concrete Technology Department
Ciudad Universitaria, Mexico

Professor A Samarin
Consultant
Sustainable Development Technological Sciences and Engineering, Australia

Professor A E Sarja
Research Professor
Technical Research Centre of Finland, Finland

Professor S P Shah
Walter P Murphy Professor or Civil Engineering
Northwestern University, USA

Professor H Sommer
Head of Research Institute
VÖZ, Austria

Professor I Soroka
Professor of Civil Engineering
National Building Research Institute, Israel

Professor M Tang
Research Professor
Nanjing University of Chemical Technology, China

Professor T Tassios
Professor
National Technical University of Athens, Greece

Professor K Tuutti
Vice President
Skanska Technik AB, Sweden

Professor T Vogel
Professor of Structural Engineering
Swiss Federal Institute of Technology ETH , Switzerland

Professor F H Wittmann
Head of Building Materials Laboratory
Swiss Federal Institute of Technology ETH , Switzerland

Professor A V Zabegayev
Head of Department RC Structures
Moscow State University of Civil Engineering, Russia

NATIONAL TECHNICAL COMMITTEE

Mr P Barber
Manager of the Scheme, The Quality Scheme for Ready Mixed Concrete

Professor A W Beeby
Professor of Structural Design, University of Leeds

Mr B V Brown
Divisional Technical Executive, Readymix (UK) Ltd.

Dr T W Broyd
Technology Development Director, W S Atkins Ltd.

Professor J H Bungey
Professor of Civil Engineering, University of Liverpool

Dr P S Chana
Director, CRIC, Imperial College of Science, Technology & Medicine

Professor J L Clarke
Principal Engineer, The Concrete Society

Dr P C Das
Group Manager, Structures Management, Highways Agency

Dr S B Desai, OBE
Principal Civil Engineer, Department of the Environment,
Transport and the Regions

Professor R K Dhir, OBE (Chairman)
Director, Concrete Technology Unit, University of Dundee

Mr C R Ecob
Director Special Services Division, Mott MacDonald Ltd.

Professor F P Glasser
University of Aberdeen

Professor T A Harrison
Technical Consultant, Quarry Products Association

Professor P C Hewlett
Director, British Board of Agrément

Professor J Innes
Director of Roads, Scottish Office

NATIONAL TECHNICAL COMMITTEE
(CONTINUED)

Mr K A L Johnson
Director, AMEC Civil Engineering Ltd.

Dr M R Jones
Senior Lecturer, Concrete Technology Unit, University of Dundee

Mr P Livesey
National Technical Services Manager, Castle Cement Ltd.

Professor A E Long
Director of School, Queens University of Belfast

Professor P S Mangat
Head of Research, Sheffield Hallam University

Mr G Masterton
Director, Babtie Group Ltd.

Professor G C Mays
Director of Civil Engineering, Cranfield University

Mr L H McCurrich
Technology Development Consultant, Fosroc Construction

Professor R S Narayanan
Partner, SB Tietz & Partners Consulting Engineers

Dr P J Nixon
Head, Centre for Concrete Construction, Building Research Establishment Ltd.

Dr W F Price
Senior Associate, Messrs Sandberg

Professor G Somerville, OBE
Director of Engineering, British Cement Association

Professor D C Spooner
Director, Materials and Standards, British Cement Association

Dr H P J Taylor
Director, Tarmac Precast Concrete Ltd.

Mr M Walker
Technical Manager, The Concrete Society

Dr R J Woodward
Senior Project Manager, Transport Research Laboratory

SUPPORTING INSTITUTIONS

American Concrete Institute, USA

American Society of Civil Engineers, USA

Australian Concrete Institute

Concrete Association of Finland

Concrete Society of Southern Africa

Concrete Society, UK

Danish Concrete Society, Denmark

Fédération de l'Industrie du Beton, France

German Concrete Association (DBV)

Hong Kong Institution of Engineers

Indian Concrete Institute

Institute of Concrete Technology, UK

Institution of Civil Engineers, UK

Instituto Brasileiro Do Concreto, Brazil

Japan Concrete Institute

Netherlands Concrete Society

New Zealand Concrete Society

Norwegian Concrete Association

Singapore Concrete Institute

Spanish Association for Structural Concrete

Swedish Concrete Association

SPONSORING ORGANISATIONS WITH EXHIBITION

AMEC Civil Engineering Ltd.

Babtie Group Ltd.

Bardon Aggregates

Blue Circle Cement

Blyth & Blyth

British Board of Agrément

British Cement Association

Building Research Establishment

Castle Cement Ltd.

Cementitious Slag Makers Association

CIMBÉTON, France

Du Pont de Nemours International S.A., Switzerland

ECC International Ltd.

Elkem Ltd. (Materials)

Fosroc International Ltd.

Grace Construction Products

HERACLES General Cement Co., Greece

John Doyle Group

Lafarge Aluminates

L M Scofield Europe Ltd.

Minelco Ltd.

Mott MacDonald Ltd.

O'Rourke Group

Ove Arup and Partners

SPONSORING ORGANISATIONS WITH EXHIBITION (CONTINUED)

Readymix (UK) Ltd.

Rugby Cement

Scottish Enterprise Tayside

Sika Ltd.

SKW - MBT Construction Chemicals

Thomas Telford Publishing Ltd.

United Kingdom Quality Ash Association

W A Fairhurst & Partners

ADDITIONAL EXHIBITORS

Christison Scientific Equipment Ltd.

CMS Pozament Limited

The Concrete Society

David Ball Group plc.

E & FN Spon

Flexcrete Ltd.

Germann Instruments A/S, Denmark

Natural Cement Distribution Limited

Palladian Publications Ltd.

Quality Scheme for Ready Mixed Concrete

UK Certification Authority for Reinforcing Steel

Wacker-Chemie GmbH, Germany

Wexham Developments

CONTENTS

LEADER PAPER

EXTENDING STRUCTURAL PERFORMANCE OF CONCRETE CONSTRUCTIONS

D Van Gemert

Catholic University Leuven

Belgium

ABSTRACT. Rehabilitation, upgrading and conservation of the vast stocks of old and degraded constructions allow financial and economical savings over new building. Extending the life span of concrete constructions is only possible if their structural performance is assured, and adapted to the expected future use of the buildings. The paper reviews recent evolutions in structural repair techniques, with emphasis on repair mortars, external steel and composite laminate strengthening, fire behavior and protection after repair. Assessment of repair performance, based on quality management of products, procedures and personnel will be presented.

Keywords: Rehabilitation, Strengthening, Performance, Quality management, Cement, Polymer, Fire protection, Personnel

Professor Dionys Van Gemert is professor of building materials science and renovation of constructions at the Department of Civil Engineering of K.U.Leuven. He is head of the Reyntjens Laboratory for Materials Testing. His research concerns repair and strengthening of constructions, deterioration and protection of building materials, concrete polymer composites.

INTRODUCTION

Public and private building owners nowadays realize the potential value of our vast stocks of old or degraded constructions as a means of providing modernized and appropriate accommodation for housing, utility and industry more quickly and at a lower cost than the alternative of new construction. Important financial savings can be obtained. The reasons for rehabilitation are numerous: the availability of existing obsolete, redundant, damaged or inappropriate constructions; the structural quality of these constructions; the shorter development period and economic advantages [1]. For rehabilitation of buildings additional advantages concerning permission issues, architecture, social aspects and others can be important.

The management of rehabilitation schemes is usually more complex than the management of new construction. The architect, the engineer and the contractor must be aware of the peculiarities and pitfalls of rehabilitation work. Without an appropriate initial building and structural survey, it is impossible to accurately assess the nature and extent of the work required and the likely cost involved. In assessing the necessary rehabilitation works, it is of vital importance to ensure that the building complies with current regulations, especially regarding the requirements for safety and serviceability of structures, as described e.g. in Eurocode 1 [2]. The 'design working life' is the assumed period for which a structure is to be used for its intended purposes with anticipated maintenance but without major repair works being necessary. Eurocode 1 indicates a design working life of 25 years for replaceable structural parts, 50 years for normal building structures and 100 years for monumental building structures, bridges and other civil engineering structures. Eurocode 1 does not explicitly cover assessment and re-design of existing structures, but it is quite clear that also for repair and rehabilitation the liability of designers and contractors extends far beyond the normal 10-years insurance period. This 10 years period particularly concerns the non-structural parts, e.g. concrete cover repairs at reinforcement corrosion and protective coatings.

Compared to this long design working life, experience with repair materials and repair techniques is still relatively short. Moreover, there is a strong and rapid evolution in the rehabilitation technologies, and in many cases structural rehabilitation techniques are based on the use of concrete-polymer composites, about which less experience is available. The long-term performance of rehabilitations is of prime importance for the involved design engineer, who is personally responsible for the safety of the building users, and that during a very long period. Long-term performance is an approximate synonym of durability: durability of a material or a building component is its ability to resist changes in its state or, in other words, of its properties. In rehabilitation, the durability problem concerns several aspects: residual performance or durability before and after rehabilitation; change of use; external reinforcement; assessing repair performance.

Extending structural performance deals with the prolongation of the working life of structures, as well as with the extension of their structural use and increase in mechanical or environmental influences on structures.

Of utmost importance in rehabilitation is real life experience, gained in case-studies. They fill the gap in theoretical knowledge about prediction of the service life or durability of materials and systems.

Evolutions in strengthening of reinforced and prestressed concrete constructions will be discussed, and criteria for the evaluation of the fire safety of these repairs are proposed. Fire hazards to externally bonded reinforcements are an obvious although not unsurmountable problem. Procedures for fire protection of the externally bonded reinforcements are presented. Besides strengthening, the load bearing capacity of concrete structures may be endangered by reinforcement corrosion. Evaluation criteria for repair mortars for concrete corrosion repair are given, and procedures for the performance assessment of structural repair materials and procedures are presented.

EVOLUTIONS IN EXTERNALLY BONDED PLATES

Plate Bonding Technique For Solving Concrete Damage

Concrete constructions can fail due to a variety of causes. Each of the damage causes, listed hereafter, covers a nearly unlimited number of specific situations, which have their own story, dramatic or amusing [3].

Calculation errors sometimes occur. If they only concern strength, they mostly remain undiscovered, because of the material safety and load factors taken into account at the design, and because of the ductility of the materials and the building structures. The situation is different when stiffness is involved. Once the total deflections exceed about 1/300 of the span, a psychological threshold is passed and people become uncomfortable in the building. In such cases the deflections must be eliminated by lifting of the concrete beam or plate. In general, due to plastic deformation of the concrete this will only be possible through cracking of the compressed zone of the concrete cross-section. As a consequence, the concrete element must be stiffened in the tension zone as well as in the compression zone [4].

Execution errors are more frequent than calculation errors. Externally bonded steel plates or fiber reinforced laminates provide a simple remedy if cracks arise in beams due to faulty overlaps of the reinforcements, if the continuity reinforcement rods in beams or plates are missing, if a supporting steel beam under a reinforced concrete plate is forgotten or if the connection between the steel beam and the concrete plates at both sides of the beam is badly designed or executed. Externally bonded reinforcements have a great flexibility to provide solutions for such damage cases.

Change of use of a building leads to other live or permanent loads on the structure, and even the structural geometry can be changed. External reinforcements can be applied to obtain considerable increase in bending and shear strength, where required.

Mechanical and other actions are a comprehensive cause of structural damage. Most striking are explosion damages and accidents [5]. Overloading is another action that may cause heavy damage to concrete structures. Environmental effects can destroy the concrete strength, e.g. freezing of balcony plates. Similar damage may be caused by fire. Creep of concrete may affect the deflections. Alkali aggregate reaction leads to destruction of concrete elements. Human mistakes often occur at the building site during construction: for the technical installations holes are drilled through columns, beams and plates with diamond cylindrical tools. If no precautions are taken, the risk of damage to the internal reinforcements is obvious. In all these cases repairs or protective treatments can be designed using epoxy bonded plates or fiber laminates [6].

Durability of External Reinforcement

Extending structural performance in time requires information about the durability of the bond against creep and environmental effects. Since the late sixties the technique is being used more and more, and its to day's popularity is the clearest proof of its durability.

However, extensive research has been done to study the long term behaviour of the glued composite: steel plate, epoxy adhesive or concrete. Pioneering work was done at EMPA-Zürich [7]. Corrosion and creep measurements were undertaken on a great number of long term exposed beams (1950 mm x 200 mm x 150 mm) strengthened with an externally bonded steel plate. This project started in 1977 and went on more than 10 years. After the exposure period no difference in static behaviour up to failure was observed. Different preparations of the bonding surfaces were compared, to check if rusting and debonding had occurred and, if so, how it could be prevented.

Additional tests were carried out in other countries [8]. Grit blasting proved to be the most appropriate preparation for both the steel and concrete bonding face. The epoxy adhesive, with and without primer, seems to have no significant influence on the corrosion behaviour. Creep of the adhesive layer is slightly higher than normal concrete creep. The creep difference increases with time although the variation is very small. It does not affect the mechanical behaviour of the steel-concrete composite element. Fatigue tests showed no fatigue failure, except at very high tensile stress variations ($\Delta\sigma = 310$ MPa) in the bonded plate. In that case plate separation from the concrete can occur, but the fatigue cracks can propagate in the concrete, as well as in the adhesive layer, depending on the specific system conditions [9].

Based on these positive experimental results, a durable external reinforcement will be obtained if the following conditions are fulfilled:

- relatively hydrophobic adhesives must be used, with a suitable rheology to ensure complete wetting of the steel surface;

- grit blasting is the minimum appropriate steel surface pre-treatment;

- stress levels in the anchoring zones must be limited, especially in fatigue loading;

- design engineers and contractors must be trained and familiar with the technique. Only certified adhesives should be used.

One of the methods for checking the durability of the glued connection under atmospheric conditions is the tear-off test on small cylinders, glued to the concrete surface. A great number of cylinders were glued on three repaired bridges over the Nete-Canal at Lier (B). Over a ten year period each time, six cylinders were tested both in the summer and winter period. After 10 years no decrease in strength could be observed [1].

The rheological behaviour of an epoxy bonded steel-concrete connection is being studied on a concrete plate reinforced with two epoxy bonded steel plates only. The loaded plate is shown in Figure 1.

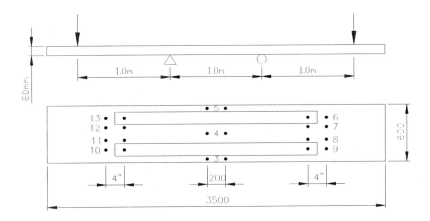

Figure 1 Externally reinforced concrete plate for outdoor exposure

The numbers indicate the different measuring bases. The loads, applied in a third-point bending test, were applied on 30-day-old concrete, whilst the glue had been cured for 7 days in laboratory conditions. The total load was 9.6 kN.

The creep test started on 05th November 1981. The rheological behaviour of the glue is evaluated by comparing the evolution of the strain on the steel-concrete bases with the evolution of the concrete-concrete strains. The evolution over 16 years is shown in Figure 2.

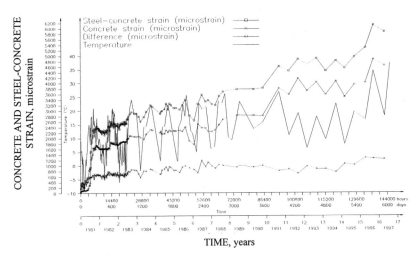

Figure 2 Evolution of temperature and strains in period 1981- 1997

Studying the temperature evolution on the same figure indicates that the first warm period had an important effect on the behaviour of the glue. An analogous variation has not been measured since. The differential creep seems to increase with time, although the variation is very small. On 12th August 1997 at 5 pm the plate structure collapsed, as shown in Figure 3.

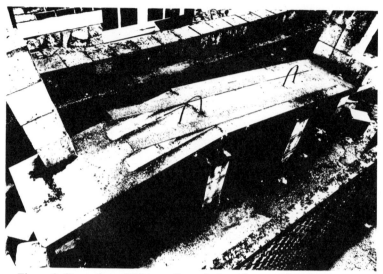

Figure 3 Strengthened Plate, collapsed after 16 years of oudoor exposure

Figure 4 Stretched surface of glue layer

The epoxy bond between steel plates and concrete failed after 15 years and 9 months of climatic exposure. The collapse took place at 17h30, at which moment the ambient temperature was still 30 °C after a very hot day with maximum temperature of 37 °C. Intense vibrations had been caused during the day by rehabilitation works on the sewer system at a distance of 2 m from the plate support. The theoretical shear stress in the glue layer was about 0.12 N/mm², but the maximum shear stress at the plate ends could amount up to 1.5 N/mm². The shear span to depth ratio was 1000/80 = 12.5 and the unplated length 400 mm. The rupture of the epoxy glue was due to viscous flow of the epoxy. This can be derived from the surface aspect of the glue after rupture, as shown in Figure 4.

The surface shows successive zones, where the glue has stretched between the steel plate and the concrete. The collapse time, several hours after the hottest moment of the day, proves that the collapse happened through an accumulation of damage in the joint, amplified by viscous flow. The fracture also happened at the plate ends, which were most exposed to direct sun radiation and heating.

Carbon Fibre Reinforced Epoxy Laminates (CFRP)

Evolution in the technology of new materials, makes it possible to replace classic steel plates by new high-grade materials. This has led to the idea of replacing the steel plates by fibre reinforced composite sheets made of unidirectional, continuous fibres such as glass, carbon and polymers bonded together with a matrix such as epoxy resin. For strengthening concrete beams, carbon fibre reinforced epoxy laminates are the most appropriate. In the early stage these laminates had to be autoclaved, which was difficult to execute in practice. The laminates were delivered on roll, Figure 5.

Since the availability of so-called prepreg laminates and pure UD-sheets, Figure 6, these difficulties have disappeared. Prepreg epoxy laminates are preimpregnated with an epoxy resin which holds the fibres together. The prepreg sheets are very flexible and can be cut easily by means of scissors. At application the prepreg sheets are impregnated again with the right ratio of epoxy resin components and the chemical reaction starts. When the first layer has hardened enough, the second layer can be applied in the same manner. Up to 10 layers can be applied. UD-sheets are completely impregnated on the site.

These prepreg and UD-sheets were first developed and produced in Japan. In Belgium the first application of on site hardening CFRP-laminates took place at the beginning of 1996 [6].

The CFRP laminates have many advantages over classic steel plates. The mechanical properties are superior. The tensile strength of the prepreg sheets is 5 to 10 times higher than that of steel, whereas the modulus of elasticity is comparable to steel. Some carbon fibre laminates reach a modulus of elasticity up to 650000 MPa. These good mechanical characteristics allow a smaller cross-section of reinforcement. Prepreg sheets are also easier to process. Since the CFRP laminates are much lighter - the density is about 3 times lower than the density of steel - the sheets can be placed with less manpower.

The prepreg sheets are available on roll which means that they are available in any length, whereas the steel plates are limited in practice to 6 metres.

Carbon fibres are very corrosion resistant. An expensive surface treatment, as for steel, is not necessary. The CFRP laminates are extremely useful in very corrosive atmospheres, such as marine and aggressive chemical atmospheres. The very thin (0.2 - 0.4 mm) laminates are very easy to apply and can be applied in cross directions without any difficulty. This is a great advantage for strengthening of two-way slabs, especially because now orthotropic behavior needs no longer be taken into account.

There are some disadvantages too. First, it is a brittle material; the carbon fibre behaves in a linear elastic way without a plastic phase up until rupture. The rupture occurs without preceding plastic deformation. The anchorage can give problems too. Due to the high stresses, the CFRP-laminates will peel off easily. The carbon fibers are longitudinally oriented and the good mechanical properties are only valid in that direction. Without special precautions it is impossible to drill a dowel hole. The fibers would be cut and no longer able to transfer forces. Special stirrup layout can offer good solutions to avoid plate end shear cracking or plate separation [11,12].

Usually it is impossible to exploit the high strength quality of the carbon fibre laminates. High stresses can only be developed at large strains and that may cause excessive stresses and strains in the concrete section. Prestressing of the laminates offers a theoretical solution to this problem, but in practice prestressing the laminates on the renovation site is mostly very difficult, not to say impossible. Finally the price for CFRP-laminates is rather high. The material is much more expensive than steel but the processing cost is much lower.

Further research should be dedicated to the fatigue resistance of the composite system: concrete/glue/laminate. Bond and crack growth, as well as delamination within the laminate must be studied. In this respect, the effect of bolts and anchorage lay-out on bond, crack growth and delamination are important aspects.

In those applications where expansion is not a problem, simple wrapping may satisfy to strengthen the elements, e.g. wrapping of columns for strengthening bridges in seismic areas. Carbon fibres may be replaced by high performance glass or other fibres. Special applications of CFRP laminates are envisaged, e.g. application for wrapping concrete columns, damaged or endangered by alkali-aggregate reaction. The column expansion of concrete damaged by alkali-aggregate reaction can be to 10 %.

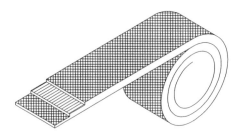

Figure 5 CFRP-sheet on roll

The wrapping laminates would not be able to allow for this expansion. Therefore research should concentrate on the production of a hybrid laminate, consisting of a 3D-sandwich fabric, as shown in Figure 7.

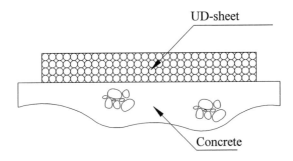

Figure 6 UD-sheet

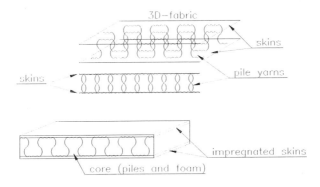

Figure 7 3D sandwich panel

Such a 3D fabric can be made to harden on the site, after application, whereby the piles can be made to stretch during hardening. In that way a stiff and strong laminate that supports the damaged element is obtained; if there is excessive expansion of the concrete section, the inner laminate skin may break, but the remaining fabric will be able to carry all loads. The core, consisting of foam and piles, must be able to absorb the swelling without damaging the outer skin of the fabric, Figure 8.

Fire Protection of Externally Bonded Reinforcements

Fire is likely to damage externally bonded reinforcements. The cold hardening glues have a glass transition temperature of about 60 to 90 °C. During a fire that temperature is easily and quickly attained if the reinforcements are exposed directly to the fire. However, the external reinforcement is usually applied to improve the degree of safety, and the total action of permanent and variable loads can be carried by the concrete structure alone, be it without any safety margin. During a fire a very low safety margin can be allowed, e.g. 1.05 to the ultimate load. If this safety is still available in the structure without external reinforcement, than one can decide to abandon a special fire protection. In most practical cases enough safety in the non-strengthened situation is still available, so that fire protection is usually unnecessary. Moreover, the investment cost for a fire protection in general is higher than the repair or strengthening cost, which makes it mostly uneconomical, compared to the risk of fire and the risk for people in the case of fire.

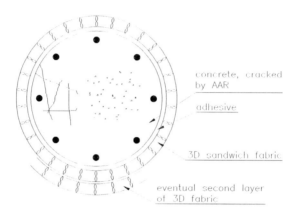

concrete, cracked by AAR

adhesive

3D sandwich fabric

eventual second layer of 3D fabric

Figure 8   Strengthening of concrete column damaged by alkali aggregate reaction

If for special reasons a fire protection is required as yet, special protective coverings are available [13]. Epoxy bonded external reinforcements can be protected by means of insulating boxes, made out of fiber reinforced silicates.

REPAIR MORTARS FOR STRUCTURAL REHABILITATION

Classification of Repair Mortars

Repair mortars for concrete are subdivided according to their use and according to the binder material.

It is necessary to distinguish between mortars that should contribute to strength and stiffness of the structure, and those who are not expected to do so. Additional requirements can be imposed regarding mechanical and physico-chemical resistance of the materials, e.g. abrasion, carbonation, diffusion.

The German regulations [14] define 4 classes of mortars:

M1: mortar appropriate to re-profile damaged zones; re-establishment of alkaline environment around reinforcing steel; sufficient bond to substrate; convenient substrate for additional protective layers.

M2: requirements for M1, plus carbonation and chloride ingress resistance; application possible under dynamic loads.

M3: repair of stressed cross-sections; requirements for
- strength and deformation, shrinkage and creep
- adhesion to reinforcing steel
- bond to substrate
- durability at elevated temperatures
- durability under sustained moist conditions.

M4: requirements for class M1, plus strength and abrasion resistance.

It is clear that each class contains mortars with different binder type, and one mortar can belong to several classes. On the basis of binder type, 3 groups of mortar are distinguished:

1. mineral mortars on cement basis, eventually with admixtures; concrete and spray concrete;
2. polymer modified mortars (PCC); spray mortar (SPCC);
3. polymer mortars.

The difference between repair mortar and concrete is the aggregate size: up to 4 mm for mortar, bigger for concrete. The maximum grain size should not exceed 1/3 of the layer thickness. Cement based mortars are applied wet in wet, and curing is necessary. A good bond remains problematic, and thin repair layers (< 20 mm) are inappropriate.

Polymer modified mortars contain a polymer resin component, added in the mix as a dispersion in water or as a dispersible powder. Recently, active functional groups such as silanes are grafted on the polymer molecular chain. They act as bond promotors because they adhere chemically to the polymer and to the aggregate surface. This improves the life time and performance under moisture conditions. Structure and composition of these mortars are close to those of the concrete substrate, avoiding major compatibility problems.

Polymer mortars are completely different from concrete, and their use is only justified if special performances are required:

1. very high chemical resistance and bond
2. fast hardening
3. curing of mineral or polymer modified mortars not possible
4. thin layers.

Typical short-term properties of the three types of mortar are given in Table 1.

More extensive discussion of repair mortars is available in literature [15, 16]. An important property concerning compatibility is shrinkage: plastic and drying shrinkage.

Shrinkage is a complex function of a multitude of factors. A non exhaustive list is given in Table 2.

Because of all these influencing factors a 0 % drying shrinkage is the target to obtain a good mortar performance, Figure 9 [16].

It is quite difficult to realise this goal, especially because the mortar passes different consistency phases during its hardening process, at which firstly expansion and afterwards shrinkage occurs. Depending on circumstances, e.g. restrained or free movements, this may result in a greater or smaller final shrinkage.

Table 1 Short term properties of repair mortars

PROPERTY	POLYMER MORTAR	POLYMER-CEMENT MORTAR	CEMENT MORTAR
	PC	PCC	CC
Compressive strength (N/mm^2)	50 - 100	30 - 60	20 - 50
Tensile strength (N/mm^2)	10 - 15	5 - 10	2 - 5
E-modulus (kN/mm^2)	10 - 20	15 - 25	20 - 30
Thermal expansion (per °C)	20 - 30 x 10^{-6}	10 - 20 x 10^{-6}	10 x 10^{-6}
Water absorption (w %)	1 - 2	0.1 - 0.5	5 - 10
Max. service temp. (°C)	40 - 80	100 - 300	> 300

Table 2 Factors inflencing the drying shrinkageof cement based mortars

MATERIAL PARAMETERS	ENVIRONMENT
- cement content	- relative air humidity
- cement type	- drying rate
- water content	- drying time
- aggregate ratio	
- admixtures	
- additives	
- porosity-compaction	
- age	
- volume/surface ratio	

Besides dimensional compatibility one has to take care about chemical compatibility as well. This concerns alkali-, C_3A-, Cl^--content etc. Electro-chemical compatibility concerns resistivity, pH. A concrete substrate containing reactive aggregates will be repaired with a low-alkali repair mortar. To avoid corrosion a repair mortar will be chose that does not make a macrocell with the substrate. Coating of a potential anodic zone will drastically change the cathode/anode surface ratio, causing accelerated corrosion. Covering the cathodic zone, where moisture and oxygen ingress must be prevented, will be th right action here.

The pull-off test is generally used to characterize the mechanical quality of the concrete surface and of the adhesion of repair materials on the concrete substrate. Not only the strength value is important, but also the type of rupture: cohesive in mortar or substrati; adhesion in interfac; mixed adhesion-cohesion. The square optimization method, developed by S. Pareek [19], is an elegant evaluation method for adhesion data.

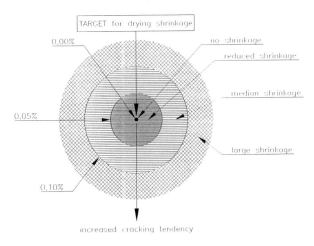

Figure 9 Targets for drying shrinkage

ASSESSMENT OF REPAIR PERFORMANCE

Polymer-modified concrete and mortars play a very important role in repair, strengthening and retrofitting of concrete structures. A lot of research has already been done in laboratories as well as by producers and ministries to set up codes of good practice to ensure durable repairs and treatments [14,17,18]. Due to a lack of mathematical or theoretical models for performance and durability predictions, most codes are based on a vast experimental testing and approval program, in which the following subdivisions are made:

- strict chemical and physical identification of material components
- mechanical and physical performance testing and approval
- qualification of personnel
- reception of materials on building site
- quality testing of executed rehabilitation work on site.

Most of these codes and procedures are in continuous evolution, and national committees are involved in preparing appropriate testing programs for special products or systems. An example are the technical agreement procedures, developed by UBAtc, the Belgian union for technical agreement in construction, who is a member of the European union UEAtc. The procedures are developed by a working group CEP-LIN-MET, in which public and private users, producers and research and control laboratories are represented. Sis sub-groups have been established, dealing with injections (SG1), structural strengthening (SG2), repair and protection of concrete and masonry (SG3), thin repair systems (SG4), special intervention techniques (SG5), quality assessment, diagnosis techniques (SG6).

These sub-groups prepare guidelines (Gxxx) for technical agreement, to be approved by the working group. Up to now about 20 guidelines have been prepared. The agreement procedure for a specific product is coordinated by executive boards. Four executive boards are already active: for coatings, thin repairs and repair mortars; for injection products; for strengthening by plate bonding; for special techniques. All the guidelines have the same structure:

- clear determination of the subject of the guideline
- terminology
- application field of the material subject to the guideline
- description of the material and application rules
- quality requirements
- description of the testing procedures
- presentation of materials for agreement testing
- quality control at the production site
- content of the agreement
- course of the agreement procedure.

The activities of the working group are in line with the European standardization efforts, coordinated by CEN. However, European standards are not yet fully available. The first standards are being published [20], and more standards are in preparation.

Besides this quality assessment procedures for materials and systems, efforts are being done to improve the quality of the personnel as well. In Belgium the certification of personnel is coordinated by the Ministries of Infrastructure and Environment, Division Control of Concrete Structures. The technicians have to prove their skill in applying the specific repair or strengthening procedure. The tests that have to be taken by the technicians are described in the ministerial document V576-B5. The examination is taken in a certified laboratory, under the supervision of a public servant.

CONCLUSIONS

Extending structural performance of concrete structures is a complex and comprehensive problem, involving technical, economical and managerial aspects. New techniques have been developed for strengthening of concrete structures by means of external steel plates or fibre-reinforced laminates. Repair mortars are now reliable components in structural rehabilitation and strengthening systems. The durability of the rehabilitation measures is being secured by a profound quality control of materials, and by an appropriate certification of the technical, executive personnel. Experience gained in case-studies and laboratory research programmes provides the calibration data and other information required to adapt and improve the codes and standards.

REFERENCES

1. VAN GEMERT D. Durability of rehabilitation, Proceedings Concrete in the service of mankind, Dundee 1996, Concrete Repair, Rehabilitation and Protection, Ed. R.K. Dhir, E & FN Spon 1996, pp. 549-558

2. CEN-ENV 1991:Eurocode 1 - Basis for design and actions on structures, CEN-Brussels, 1994

3. VAN GEMERT D. Design, application and durability of plate bonding technique, Proceedings Concrete in the service of mankind, Dundee 1996, Concrete Repair, Rehabilitation and Protection, Ed. R.K. Dhir, E & FN Spon 1996, pp. 559-570

4. VAN GEMERT D. Repairing of concrete structures by externally bonded steel plates, ICP/RILEM/IBK International Symposium 1981, Prague, Plactics in Material and Structural Engineering, pp. 519-526

5. VAN GEMERT D. Evolutions in design and application of the plate bonding technique, Symposium on Structural Repairs/Strengthening by the Plate Bonding Technique, Sheffield, G.B., September 6-7, 1990

6. VAN GEMERT D. Evolutions nouvelles en matière d'armatures collées pour le renforcement du bétonProceedings Conference 'Armatures défaillantes! Que faire?', Liège 20.03.96, 10p.

7. LADNER, M. Belastungsversuche an 12 fünfjärigen Stahlbetonbalken mit aufgeklebter Bewehrung, EMPA Bericht nr 37'846/4, 1983, Zürich, Switzerland

8. HOBBS B., JONES R., SWAMY N., ROBERTS M. Accelerated tests to assess the durability of bonded plate systems, Int. Seminar Structural Repairs/Strengthening by the plate bonding technique, Sheffield, 1990

9. HANKERS CH. Strengthening of structural concrete members by bonded steel plates. Fatigue behavior and fatigue design, Int. Seminar Structural Repairs/Strengthening by the plate bonding technique, Sheffield, 1990

10. VAN GEMERT D. Special design aspects of adhesive bonding plates, ACI SP 165, Ed. N. Swamy and R. Gaul, 1996, pp. 25-42

11. TÄLJSTEN B. Plate bonding. Strengthening of existing concrete structures with epoxy bonded plates of steel or fibre reinforced plastics, Doctoral Thesis, Luleå University of Technology, Sweden, 1994

12. VAN GEMERT D., AHMED O., BROSENS K. Anchoring of externally bonded CFRP reinforcement, Int. Congress Creating with Concrete, 1999

13. ETERNIT-PROMAT Fire prevention in constructions. Manual 91 (in Dutch), Eternit Group 1991

14. ZTV-SIB 90 Zusätzliche Technische Vertragsbedingungen und Richtlinien für Schutz und Instandsetzung von Betonbauteilen, Verkehrsblatt-Verlag, Dortmund 1990

15. VAN GEMERT D. Repair of concrete surfaces (in Dutch), Innovation in construction, K.VIV-TI Antwerp, 1996, pp. 167-178

16. EMMONS P., MCDONALD J., VAYSBURD A. Some compatibility problems in repair of concrete structures - a fresh look, Materials Science and Restoration, vol 1, pp. 836-857, Expert-Verlag 1992

17. OHAMA Y. Handbook of polymer-modified concrete and mortars, Noyes Publications, New Jersey, USA, 1995

18. IBAC-Aachen Schutz und Instandsetzung von Betonbauteilen. Die Neue DafStb-Richtlinie, RWTH Aachen, 04.10.1989

19. PAREEK S. Improvement in adhesion of polymeric repair and finish materials for reinforced concrete structures, PhD thesis, Nihon University, Koriyama, 1993

20. CEN EN1504 Products and systems for the protection and repair of concrete structures - Definitions, requirements, quality control and evaluation of conformity. Part 9, July 1997. Part 1, January 1998. Parts 2 - 8 in preparation.

THEME ONE:

MATERIALS DEVELOPMENT

DEVELOPMENTS IN CEMENT-BASED COMPOSITES

A M Brandt

Polish Academy of Sciences

L Kucharska

Technical University

Poland

ABSTRACT. In this paper recent developments in concrete-like materials are presented and discussed. New components and techniques, and outstanding properties are outlined in view of new requirements and their application in structures is briefly reviewed. More attention is paid to the problems considered in basic and applied research, and these are related to advances in concretes which are transformed into composite materials of high technology.

Keywords: Concrete components, Admixtures, Pozzolans, Microfillers, Microreinforcement, DSP, MDF, HSC, HPC, RPC, HPFRCC.

Professor Andrzej M Brandt is Head of the Strain Fields Department of the Institute of Fundamental Technological Research (IFTR), Polish Academy of Sciences, Warsaw, Poland. His main research interests are in cement based composites, their mechanical behaviour and performance. Dr. A.M. Brandt is the author of several papers and books on optimization and mechanics of structures and of materials. He is a consulting member of ACI 544 Committee (Fiber Reinforced Concrete).

Professor Leokadia Kucharska is Head of Building Materials Department at the Technical University, Wroclaw, Poland. Her background is chemistry and she is interested in relations between composition, rheology, structure and properties of brittle engineering materials. Recently, she has been involved in research in High Performance Concretes and in cement based fibre reinforced composites. Dr. L. Kucharska has published numerous papers in the field of ceramics and cement composites.

INTRODUCTION

Development in science and technology is composed of processes that may be represented by curves combined with continuous and discontinuous sections. During each period of continuous and relatively slow development there is an agglomeration of small discoveries, improvements and new steps, either in theoretical methods and solutions or analytical approaches and experimental methods. After such a period often a considerable jump occurs, caused either by a new and important discovery or by an outstanding recapitulation of all past small advances into a new and creative image. Such jumps in the civil engineering are related to new materials and technological achievements or to revolutionary analytical methods. Quite often a break-through is based on a multidisciplinary progress achieved in other fields and successfully transferred into the civil engineering.

There are several examples of such discontinuities. In a few last centuries we note the introduction of iron and steel to the engineering constructions, discovery of Portland cement, methods of execution and calculation of reinforced and prestressed structures, etc. On the more theoretical side there are: application of the rational mechanics in design of structures, development of the mechanics of solids and the theory of probability, more recently the finite element method with all its refinements, and also application of the fracture mechanics.

In our century and in the field of concrete-like materials several important developments occurred: discovery of the relation between water/cement ratio and the concrete strength, introduction of air entrainment and its importance for the durability of concrete, invention of admixtures and particularly these which allow to reduce water demand of the fresh mix. Certainly other important innovations may be added to that list.

If we concentrate our attention on last few decades, we may conclude that new and very significant steps were made: large application of several types of admixtures, use of microreinforcements and microfillers, new methods of testing and measurement, both in laboratories and in situ, utilization of by-products and industrial wastes, etc. Probably we are at present in one of such periods of discontinuity that are fascinating and attractive. That opinion was presented by Frohnsdorff and Skalny [1], Figure 1.

It is a privilege of historians to classify the past into well organized periods and to characterize them with short and elegant descriptions. It is however more difficult to recognize the present developments and to evaluate them correctly.

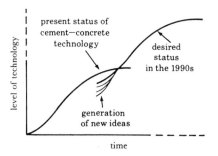

Figure 1 Discontinuities in the knowledge transfer on practice, after [1]

By its nature, the domain of building and civil engineering has a conservative character, related to such stable factors as human needs of comfort, safety and durability, the environmental influences, main raw materials, etc. This results in considerable delays between actual appearance of new products or methods and their large application which just creates a new technological achievement.

The aim of the paper is to discuss the most important developments in the field of the structural concrete mainly from the viewpoint of their impact on concrete structures. Some kinds of concrete like lightweight concretes are not considered here in detail because their developments in last few years is perhaps less significant. The admixtures that play a considerable role in developments in concrete are not described in a separate chapter but are mentioned at many places. New kinds of concrete reinforcement like non-metallic tendons, and laminates, epoxy-coated and corrosion resistant steel bars, as well as methods of cathodic protection are considered as out of scope of the paper.

CONCRETE AS A MAIN BUILDING MATERIAL

For mankind, concrete is the most important building material. Its consumption is increasing and already in the mid 1990s it reached one cubic meter or 2.5t per head as average in the world and the production of cement is approaching 10^9 tons per year. The increasing demand for housing in the developing countries, intensive reconstruction of infrastructure after the World War II and after other multiple conflicts, construction of roads and bridges related to the tremendous increase of traffic in most countries in the world, these are only some factors which increased the scope of various applications and the demand for Portland cement and cement-based building materials.

We are aware of the fact that quite often the concrete is not as strong and durable as it is possible due to its components and internal structure: with the same components and practically at the same price the concrete can be much better! That was the starting point in early 1980s when an important break-through in the concrete technology was initiated, aimed at strong and durable concrete composites.

STRONG MATRICES AND CONCRETES

Already in ancient Greece Titus Lucretius Carus (95-55 BC) wrote: "The more vacuum a thing contains within it, the more readily it yields...". The development of strong composites is based on increasing their density and reducing porosity. A few well known abbreviations mark important steps in the development of cement-based composites; these are DSP, MDF, HSC and HPC, and finally RPC.

Very high strength cement matrices were already obtained nearly seventy years ago. In the laboratory of the Lone Star Cement Corporation in USA in 1930, under direction of Duff A. Abrams, the compressive strength of cement paste equal to 276 MPa was achieved at 28 days, Powers [2]. The paste was of a very low water/cement ratio equal to 0.08 and was subjected to high pressure. This amazing event had no particular influence on contemporary practice and up to the 1970s the strength of concrete only rarely exceeded 30 to 40 MPa, because there were no methods neither additional components to increase the strength in practical conditions in situ.

Next record in strength was only obtained in the 1980s with a new material called DSP - Densified with Small Particles, Bache [3], Hjorth [4]. These were called also Densified Systems containing homogeneously arranged ultrafine particles. The composition of the particles in right proportions ensured dense distribution of grains to fill the space: smaller particles entered into interstitial spaces between Portland cement grains, that in large part are non-hydrated. The pozzolanity of both admixtures increased the material strength with its age. DSP are characterized by a compressive strength up to 250 MPa.

Another material of strength around 400 MPa was MDF - Macro-Defect Free paste, Rendall, Birchall [5], where the rheology of the fresh mix was modified by addition of a water soluble polymer of around 7% by cement mass. High energy shear mixing and low pressure (about 5 MPa) during hydration enabled cement particles to be rearranged and their closer packing was obtained. The problems of the affinity of polymers to water and swelling are still examined.

The experience gained with these special matrices and the general application of superplasticizers created a new kind of concrete called HSC - High Strength Concrete. That structural material was produced in the late 1980s in several countries at normal conditions of a site. Better quality of aggregate, careful selection of the grain distribution down to the small diameters and low water/cement ratio resulted in the compressive strength at 28 days exceeding 60 MPa without special microfillers. Addition of silica fume is considered as necessary only for the strength over 100 MPa or so.

Next, a different concept was introduced and called HPC - High Performance Concrete. As translated into technical terms of cement-based materials its meaning is much larger than just the high compressive strength. As main features of HPC the following are usually mentioned: excellent workability of the fresh mix, high density and high early strength. That list may be considerably extended, because in fact HPC means a concrete designed and executed just for the purpose. It was observed that if high workability and improved density are ensured, then the high strength and other mechanical properties are obtained without additional effort and expense, Malier [6].

In the list of conditions that should be satisfied in view to obtain HPC the low w/c ratio is certainly the most important. This is obtained by decreasing the volume of water and related increase of the binder content. To maintain and improve the workability of the fresh mix the use of platicizers (early 1980s), and later superplasticizers, is prerequisite and their universal application since 1980s was a major step forward, Mindess [7]. The admixtures became important components of concretes nearly for all purposes. They improve the flowability, regulate time of setting, hardening and shrinkage, improve the impermeability and frost resistance, etc. [8]. A competent design of mixture proportions, frequently using computer-assisted optimization procedures are needed for multi-components concretes.

At present, the compressive strength of VHPC equal to 140 MPa can be obtained routinely in several countries. For that quality of concretes the use of microfillers like silica fume is practically necessary. The evolution of such notions like high strength and high performance is schematically shown in Table 1, Brandt [9], partly using ACI specifications. The discontinuity of this process is presented in Figure 2.

Table 1 Evolution of the notion of high compressive strength concrete, after [9]

QUALIFICATION OF CONCRETE	YEARS AND COMPRESSIVE STRENGTH IN MPa						
	1850	1910	1950	1979	1981	1998	2000+
OC	8	15	30	<40	<60	<60	?
HPC				>40	>60	>60	?
VHPC					>120	>120	?

OC - Ordinary Concrete, VHPC - Very High Performance Concrete

For practical reasons it is important, that HPC is not basically different from concretes that are used since years. Therefore, neither its higher unite price nor some necessary additional care in design and execution delayed the introduction HPC into the field of building and civil engineering.

On the contrary, as it was proved by several examples, the increased durability alone is able to repay all additional expenses related to better components and workmanship.

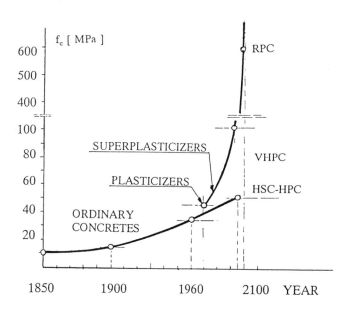

Figure 2 Evolution of concrete compressive strength

In the field of HPC there are still several problems that wait for intensive research and new solutions. These are their increased brittleness, autogenous shrinkage, special requirements for intensive curing, behaviour at elevated temperature, etc. Fracture properties, crack growth mechanisms and the analysis of fractal dimension of the fractured surfaces are also subject of many investigations, e.g. Zhou et at. [10]. The influence of the interfacial zone between cement paste and aggregate grains in HPC is more important than in ordinary concretes, Diamond, Huang [11], Kucharska [12].

Recent development is the ultra-high strength ductile concrete (Reactive Powder Concrete - RPC) of 200 - 800 MPa compressive strength. Its concept is the homogeneous composition and granular compacted density. The basic principles are:

- elimination of coarse aggregate (max. 0.6 mm),
- application of pressure before and during setting,
- post-set heating to enhance microstructure,
- introduction of fine steel fibres to improve ductility.

RPC may be applied for replacement of steel in the mechanical parts, nuclear waste storage facilities, military structures and equipment. Very high compressive strength between 200 and 800 MPa and fracture energy up to $40kJ^{-2}$ are obtained, Richard, Cheyrezy [13].

MICRO-REINFORCEMENT

Relatively high brittleness of very dense composites can be controlled not only by the main reinforcement of the structural elements but also by dispersed fibres. Application of short fibres of various materials as micro-reinforcement for concretes and mortars is since years a normal practice in building and civil engineering structures. Recent developments concern steel micro-fibres, polypropylene fibres, pitch-based carbon fibres and new techniques in improvement the durability of alkali-resistant (AR) glass fibres in cement paste.

When the fibres are efficient enough to increase the bearing capacity of an element after first cracking, then a quasi-strain hardening is observed, Fig. 3. That kind of composites is called High Performance Fibre Reinforced Cement Composites - HPFRCC, Naaman, Reinhardt, [14, 15].

The micro-fibres of a few millimeters in length are more efficient because of better dispersion in the matrix. Great number of micro-fibres in a unite volume enable to control opening and propagation of the micro-cracks. In Table 2 approximate data are shown calculated after average fibre dimensions to present how densely the micro-fibres are distributed. The increase of toughness is the most important effect of dispersed fibres in the brittle cement matrix.

Corrosion of the glass fibres and embrittlement are reduced not only by the improvement of the fibre composition but also by modification of the matrix with various admixtures to reduce its alkalinity (metakaolin, fly ash, bentonit, etc.). The incorporation of metakaolin up to 15% in cement paste not only reduces its alkalinity but also leads to refinement of the pore structure: the percentage of pores below 20nm is increased, Khatib, Wild [16].

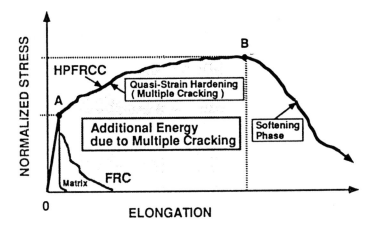

Figure 3   Quasi-strain hardening curve of HPFRCC, after [15]

The use of polypropylene fibres has increased in last years. The fibres are perfectly resistant to corrosion and are mainly applied to control cracking of cement matrix due to early shrinkage. The Young modulus of concretes at early age is similar to that of the fibres and their control of microcracks is efficient.

Carbon fibres (CF) are excellent reinforcement for thin elements with cement paste matrix. They are resistant to all types of chemical attack and to high temperature. Because of small dimensions, their dispersion allows to control microcracking, Kucharska, Brandt [17].

Two recent developments in the fibrous reinforcement merit particular attention. Slurry Infiltrated Fibre Concrete (SIFCON) is manufactured with high volume of steel fibres.

Table 2   Number N of single fibres in 1cm³ of the matrix

Type Of Fibres	Pitch Carbon	Pan Carbon	Asbestos	Polypr-opylene	Pva	Steel Micro Fibres	Steel Ordinary Fibres
diameter μm	14.5-18	6-10	0.02-25	18	12-41	150	250-600
length m	3-12	3-6	1-5	10-25	4-12	3-6	25-60
number N *)	20200	118000	3.18×10^8	3300	3200	94	2.65

PAN - polyacrylonitrile, PVA - polyvinyl alcohol
*) Number of fibres calculated for $V_f = 1\%$ and average dimensions of a single fibre

The fibres are lied in a form and then infiltrated with very liquid cement slurry. In SIMCON (Slurry Infiltrated Mat Concrete) non-woven mats are used instead of single fibres. Because of very strong and dispersed reinforcement the composites exhibit considerable toughness and impact strength. Various kinds of the fibre reinforcement are applied commonly in industrial floors, bridge and road pavements, claddings, foundations for heavy machinery, etc.

POLYMER CONCRETES

Cement based composites are improved by various polymers to enhance their ductility, impermeability and adherence that is important in repair works. There are several kinds of concretes with polymers as components and not only as admixtures. These are: Polymer Concretes (PC) where polymer is a binder, Polymer Impregnated Concretes (PIC) which are ordinary cement concretes impregnated with polymers, and Polymer Cement Concretes (PCC) with two binders composed together. Polymer concretes are attracting attention mostly for repair works of concrete structures and as overlays for special purposes, e.g., requiring long term durability in corrosive environment, Fowler [18]. In the development of the polymer concretes a considerable progress achieved in the polymers themselves is used.

PROPERTIES OF THE FRESH MIX

The traditional test of workability by the Abram's is still used in situ, but the complete rheological tests give more relevant information. In last few years the importance of the properties of the fresh mix was better understood. Two-parameter Bingham model is universally applied for characterization of the behaviour of the fresh cement mix, Tattersall [19]. This simplified model is used with a few types of rheometers for determination of these parameters: yield stress and plastic viscosity. By the rheological tests the compatibility between cements and admixtures can be verified, the amount of admixtures can be established, the optimum duration of the workability and the sequence of mixing can be determined. The test results allow not only to estimate whether the concrete is adapted for the purpose in the sense of its flowability, pumpability, etc., but also predictions of its properties after hardening are possible.

Recently de Larrard [20] proposed a more sophisticated three-parameter model in which relation between the shear stress and shear rate is nonlinear. The fluidity of the fresh mix is controlled not only by w/c ratio and superplasticizers. The investigations have shown that the spherical shape of cement particles as well as their fineness have considerable influence, Tanaka et al. [21].

USE OF BY-PRODUCTS AND WASTES AS CONCRETE COMPONENTS

There are two main reasons that various kinds of by-products and waste materials are used either as pozzolanic materials in production of blended cements and as admixtures or for production of artificial aggregates.

In all countries there is an urgent need to find use for the increasing amount of these materials agglomerated in industrial regions. One of the important use is in Portland cement concretes, without compromising their performance for certain applications, e.g. the blended cements can contain up to 50% ground-granulated blast-furnace slag (GGBS). Furthermore, several kinds of hazardous and toxic wastes, with heavy metals or having low level radioactivity can be safely immobilized in concrete structures, e.g. dams, roads, embankments, etc.

Due to the considerable growth in demand of all kinds of concrete, already now in certain regions of the world good quality of aggregate is exhausted and this situation will be probably intensified in the future. The demolished concrete structures are also re-used in the form of full value aggregate.

The presence of supplementary cementing materials may have a positive influence on the hydration kinetics of cement, on the workability of the fresh mix and finally on the physical properties of mortar and concrete. For instance, blended cements with slag and fly ash concretes have many advantages over Portland cement (PC), Dhir, Matthews [22]. Some of these advantages are higher long term concrete strength, lower cost, better durability, lower heat exhaustion, etc. The resistance of concretes is improved against chemical attacks and freezing/thawing cycles and repressed alkali-aggregate reaction, Jiang et al. [23].

The replacement of cement (15-25%) by fly ash has a beneficial effect on the tensile strength of concrete, thus decreasing its brittleness. The problems with the proper applications are not solved yet for the ashes from fluidized bed combustion. It is noticeable that some materials, though not in compliance with existing specifications, can still be efficiently used so long as they are properly considered in design of the mixture proportions and adequately treated in technological operations.

There is enough knowledge and experience to produce concretes with natural and artificial pozzolans equal to, or better than, those using pure Portland cement. The double and triple blended cements can be used also in most kinds of the high performance concretes, provided that the quality of all components and an appropriate mixture composition are ensured. The variety of artificial pozzolans is increasing. Certain by-products became highly demanded admixtures for high performance concretes, e.g. silica fume, first used already in 1970. Its influence is a combination of the filler and the pozzolanic effect and may be summed up as improved rheological behaviour (ball-rolling), homogenization of the microstructure, segmentation of pores and improved the interfacial transition zone (ITZ) with aggregate and reinforcing steel. As the effects the following may be mentioned: reduced plastic shrinkage, decreased permeability and improved mechanical properties.

Many researchers observed that silica fume reduces the temperature rise in blended cement paste, slightly accelerating the peak temperature. The long-term performance of concretes with silica fume was proved on structures exposed to severe conditions during over 15 years, Lachemi et al. [24].

Significant influence of silica fume on the concrete resistance to the ingress of chloride ions is observed thus the corrosion rate of the reinforcing steel can be considerably reduced.

Because the price of silica fume is steadily increasing and its availability on the market will inevitably decrease, its replacement is looked for; in certain countries burnt rice husk is used, and also ashes, zeolites, ground limestone and metakaolinite.

DEVELOPMENTS IN TECHNOLOGY

Several new technological methods have appeared in the last few years and for sake of brevity only a few of them are here mentioned. The rapid strength development of concrete can greatly facilitate construction operations, precasting procedures and structural repairs. One of the developing techniques is the microwave curing. It is superior to conventional thermal curing in two important aspects:

- very high early strength can be obtained,
- there is only little or no strength deterioration in the longer term.

Increase of rate of hydration and hardening of cement paste is realized in various ways, e.g. microwave curing, special admixtures and microcements (with $d_{50} = 6$ or 12 μm). Microwave curing is also applied for accelerate curing of glass fibre reinforced cement elements with up to 75% of cement replaced with metakaolin, Leung [25]

For concrete dams, pavements for airfields, streets and parking lots, etc. the Roller Compacted Concrete (RCC) is a special technology. This is an extremely dry Portland cement concrete designed for zero-slump and compacted by heavy vibratory rolling. Adequate compaction properties are ensured by appropriate sieve distribution of the aggregate grains. In this technology considerable economies in time and cost are obtained with respect to conventional pavements, together with high strength and long term durability.

The technique of shotcreting mortar or fibre mortar is applied frequently for repair concrete structures. Depending on the volume of work and local conditions dry-mix or wet-mix technique are used. Addition of short steel or polypropylene fibres improves considerably the quality of repair. When optimum mixture proportion is prepared and other parameters selected correctly, then the layers 50-70 mm thick can be applied in a single passage with the rebound reduced down to 25%.

For works in tunnels or other difficult sites the shotcreting is executed by robots and workers may be kept outside the dangerous zone. Normally a layer of shotcrete does not require any special finition to ensure perfect durability.

Two other kinds of technologies are developed in last few years. Self-leveling concrete for industrial floors does not need any vibration or other operation to obtain plane and horizontal surface. The mixture proportions are designed in such a way that high fluidity is obtained without segregation. For self-cured concretes special compounds with polymers and oils are included into fresh mix. These products rise up to the surface at the beginning of the cement setting and seal the surface against evaporation.

Verification of compatibility between all components is an important question that should be answered when a concrete composition is designed.

This is particularly necessary now when a large variety of cements and admixtures are available on the market. Here the rheological methods mentioned before are very useful.

DURABILITY OF CONCRETE

The concrete is produced in nearly all countries in the world but often with low quality aggregates and without adequate care. In the past the high knowledge in concrete technology was used only for special structures: high-rise buildings and long-span bridges. In the ordinary structures low quality material with poor durability was used and their resistance against various types of corrosive agents was insufficient. As a result, the structures deteriorated rapidly and required basic repairs often after a few years of exploitation. Already in 1970s when the structures built shortly after the World War II reached at the age of 30 years or less, it became clear that the national and regional budgets were overloaded by the costs of repairs and rehabilitations. Such situations occurred even in the countries with high level of technical and technological development like USA and Japan. It is also now an important problem at a universal scale to transform the concrete into a high quality durable material.

High strength does not necessarily means adequate durability of the concrete, as it is still assumed by many practitioners who rely only on the 28-day compressive strength as on the main if not unique mechanical parameter. The concrete is a composite material characterized by a multitude of parameters that are only partly interrelated. The durability should be considered as a holistic criterion, Mehta [26].

The design of concrete structures for adequate durability under loads and environmental agents requires cooperation of specialists from at least two fields: chemistry and mechanics. The hydration of cement and other cementitious components, development of the microstructure and evolution of the mechanical properties are interrelated. The final product has to satisfy all requirements, and the most important is its durability.

There are several processes by which the deterioration of concrete originate and progress up to advanced stages that require immediate repair and retrofit. The corrosion of reinforcing steel is, by far, the main cause of the premature deterioration of highways, roads, bridge decks and marine structures. It is initiated by carbonation of the concrete external layer and by penetration of the chloride ions.

Rate of transport of the ions and other species through the pore systems depends among the other on the concrete structure. The concrete microstructure and its optimization and modelling is one of the most important area of research in last years, Zia [27]. The density of concretes is increased by the application of high rate water reducers and the continuous sieve distribution is looked for from zero up to the maximum grains. Dense packing of various particles in the microstructure results in high strength and low permeability of the concrete systems.

Modern Portland cements are characterized by relatively high heat of hydration at early ages that is related to quick increase of strength.

This may result in additional microcracking due to thermal gradients. The system of cracks and microcracks is also a way through which external agents enter inside and cause corrosion.

Another cause of deterioration of concrete structures in many countries is the Alkali-Aggregate Reaction (AAR), Swamy [28], Idorn [29]. The chemical processes that develop between reactive aggregate grains and alkalis in the cement paste, when the thermal and hygral conditions are favourable, result in production of a gel able to swell that may have immediate mechanical effects in the form of cracks. All these phenomena are random and may appear in new and old structures, like bridge beams, railway ties, retaining walls, etc. The cracks cause disintegration of the structural elements, also by inducing corrosion of reinforcing steel. In less frequent cases, the AAR results in systems of the pop-outs on external faces of concrete structures, e.g., industrial floors.

The preventive measures, methods of diagnosis of aggregate and modelling of the process are the object of intensive research in many countries, because in last years the AAR was observed all over the world, Bournazel, Moranville [30].

Important cause of deterioration of concrete structures is the frost and freeze/thaw cycles. In many countries the number of cycles in a year can reach hundreds and the concrete surface is exposed to de-icing salts. Application of air-entraining agents and increase of the density of concrete microstructure are two main directions of defense. The problem whether HPC and VHPC do require air-entrainment is frequently submitted to tests and analyses. The opinions are dispersed and depend on regional and national standards, but also on the climate. Probably for out-door structures exposed to very low temperature all preventive measures should be applied at a time. There are also doubts how long the air-entrained system of pores remain in the hardened concrete.

All particular requirements like enhanced resistance sulfate attack, against shocks, in hot climate, in marine environment, etc., can be to some extent satisfied by high performance concretes designed for special use.

MATERIALS FOR REPAIR AND REHABILITATION OF CONCRETE STRUCTURES

The repair of concrete structures is an important problem because in most countries the large part of existing structures need repair, retrofit or upgrading. Various processes that develop in both materials: repair and old one, may cause chemical, thermal and mechanical effects with negative results on the durability of the repair. Outstanding progress observed in a few last years in the repair methods is related to the achievements in new materials and technologies. Application of the composite materials: fibre reinforced cements (FRC), slurry infiltrated concretes (SIFCON and SIMCON), polymer concretes, etc., with appropriate admixtures to reduce shrinkage and increase adherence, allows to overcome most of factors that endanger the durability. Quite frequently the application of the technique of shotcreting, sometimes with the help of robots and full automatization, is a viable solution.

Rehabilitation and strengthening is required for structures that are not damaged, but their load bearing capacity and/or aspect do not satisfy present requirements.

The programs for structural strengthening are often set up for large number of structures in a region because of new regulations. In many countries large programs are initiated based on new methods, materials and techniques used for repair and strengthening of concrete structures.

Concrete is also used for repair of asphalt pavements (roads and runways) in the form of so called "white topping", i.e. 70-100 mm deep layers cast directly on old asphalt.

ECONOMY OF CONCRETE AND CONCRETE STRUCTURES

At present, in the calculations of cost of the concrete and concrete structures more often than before the life-cycle cost is considered. Although the initial cost and the cost of maintenance and repairs during service are often covered from different sources, all money is paid by the society in general.

The initial cost of a HPC is usually higher than that of an ordinary concrete: better components, admixtures like superplasticizers and silica fume, careful execution with better curing and tight control are the elements of some additional expenses. On the other hand smaller mass of concrete and its higher strength give appreciable benefits already in the sense of initial cost of the structure, e.g. reduced cross-sections of columns allow for larger usable space in buildings, with stronger concrete the bridge deck may be composed of a reduced number of girders, longer bridge spans require a smaller number of supports, etc.

The HPC is however more durable, and this is the main benefit, especially for structures exposed to climatic influences or to corrosive agents. Durability means less frequent repairs and smaller probability that the replacement will be necessary already after few years. All this save money for the owner of the structure. Professional engineers got in last few years a large variety of components and techniques to make strong and durable concretes. Furthermore, there is always a choice: either to select high quality components and careful execution or apply more expensive admixtures like silica fume and chemicals. Probably the intermediary solutions are to be selected using optimization procedures with the most important criterion that is the cost of the structure, Brandt [31]. There are new techniques applied for concrete mixture proportions, e.g., Artificial Neural Networks, which are particularly useful in the case of the design of the mixture proportions with a large number of components.

The life-cycle costing of structures is the main reason that the high performance concretes with their properties designed for the needs are produced in increasing rate in the world, Neville, Aïtcin [32].

CONCLUDING REMARKS

There is no doubt that the cement-based materials will be further developed and in that respect the following questions may be asked:

1. Will it be possible to decrease significantly the energy consumption in the production of cements, e.g. by application of new catalyzers for reactions in the solid state ?

2. Whether the chemical bonds can replace the physical ones to improve considerably the properties of hardened concrete ?

3. To what extent the present use of non-metallic reinforcement will be further developed ?

4. What advanced and unexpected applications will appear ?

The list of questions may be continued and only the future will answer them.

Outstanding structures such as The Troll Offshore Platform in Norway, the Petronas Towers in Kuala Lumpur (Malaysia) and bridges and tunnels across Danish Straits demostrate the advanced concrete technology that will be required to satisfy the needs of the next millennium.

ACKNOWLEDGMENT

In this paper, information from a large number of papers and books was analyzed and introduced to the text. For obvious reasons it is possible to mention only a small part of authors and publications where new and important data are given. The authors ask all researchers who are not shown in the list of references to understand that situation.

REFERENCES

1. FROHNSDORFF, G., SKALNY, J., Cement in the 1990s: challenges and possibilities. Phil. Trans. Royal Soc., A 310, London 1983, 17-29

2. POWERS, T.C. A discussion of cement hydration in relation to the curing of concrete. Proc. of Highway Research Board, 27th Ann.Meeting, Washington D.C.,

3. BACHE, H.H. Densified cement-ultrafine particle-base materials. Proc. 2nd Int. Conf. on Superplasticizers in Concrete, Ottawa 1981

4. HJORTH, B.L. Development and application of high-density cement-based materials. Phil. Trans. Royal Soc., A310, London 1983, 167-173

5. KENDALL, K., BIRCHALL, J.G. Porosity and its relationship to the strength of hydraulic cement pastes. Proc. Symp. Very High Strength Concrete-Based Materials, 42, Material Research Society, Pittsburgh 1983, 153-158

6. MALIER, Y. ed. Les bétons à hautes performances. Presses de l'ENPC, 1992, 673pp

7. MINDESS, S. Materials selection, proportioning and quality control. High Performance Concretes and Applications, S.P.Shah and S.H.Ahmad,. E.Arnold, London 1994, 1-25

8. RILEM. Draft State-of-the-Art Report on Admixtures in High-Performance Concrete. RILEM TC 158-AHC. Materials and Structures, suppl. March 1997, 47-53

9. BRANDT, A.M. Cement-Based Composites. Materials, Mechanical Properties and Performance. E&FN Spon/Chapman&Hall, London 1995, 470pp

10. ZHOU, F.P., BARR, B.I.G., LYDON, F.D. Fracture properties of high strength concrete with varying silica fume content and aggregates. Cem.&Concr.Res., 25, 3 1995, 543-552

11. DIAMOND, S., HUANG, J. The interfacial transition zone: reality or myth? Proc. 2nd RILEM Conf. on the Interfacial Transition Zone in Cementitious Composites, Technion, Haifa 1998, 1-40

12. KUCHARSKA, L. W/C ratio as an indication of the influence of ITZ on the mechanical properties of ordinary concrete and HPC (in Polish), Proc. 2nd Conf. on Materials in Civil Engineering MATBUD'98, Cracow 1998, 241-250

13. RICHARD, P., CHEYREZY, M.H. Composition of reactive powder concretes.Cem.&Concr.Res., 25, 7, 1995, 1501-1511

14. NAAMAN, A., REINHARDT, H.W., eds. High Performance Fibre Reinforced Cement Composites - HPFRCC. E & FN Spon, London 1992, 565pp

15. NAAMAN, A., REINHARDT, H.W., eds. High Performance Fibre Reinforced Cement Composites 2 (HPFRCC 2). E & FN Spon, London 1996, 502pp

16. KHATIB, J.M., WILD, S. Pore size distribution of metakaolin paste. Cem.&Concr.Res., 26, 10, 1996, 1545-1553

17. KUCHARSKA, L., BRANDT, A.M. Pitch-based carbon fibres reinforced cement composites, Arch. of Civ. Eng., 42, 2, 1997, 165-187

18. FOWLER, D.W. Overview of the use of polymers and polymer fibres in concrete. Third Southern African Conference on Polymers in Concrete, Johannesburg 1997, 8pp

19. TATTERSALL, G.H. Workability and quality-control of concrete. E & FN Spon, London 1991

20. De LARRARD, F., FERRARIS, C.F. Rhéologie du béton frais remanié. Bull.des Lab. des Ponts et Ch., no 213, 1998, 73-89

21. TANAKA, I., SUZUKI, N., ONO, Y., KOISHI, M. Fluidity of spherical cement and mechanism for creating high fluidity. Cem.&Concr.Res., 28, 1, 1998, 63-74

22. DHIR, R.K., MATTHEWS, J.D. Durability of PFA concrete. International Cement Review, May 1992, 75-81

23. JIANG, W., SILSBEE, M.R., ROY, D.M. Similarities and differences of microstructure and macro-properties between Portland and blended cement. Cem.&Concr.Res., 27, 10, 1997, 1501-1511

24. LACHEMI, M., LI, G., TAGNIT-HAMOU, A., AÏTCIN, P.C. Long-term performance of silica fume concretes. Concr. Int., January 1998, 59-65

25. LEUNG, C.K.Y., THANAKORN PHEERAPHAN. Determination of optimal process for microwave curing of concrete. Cem.&Concr.Res., 27, 3, 1997, 463-472

26. MEHTA, P.K. Durability - critical issues for the future. Concr. Int., Judy 1997, 27-33

27. ZIA, P. High-performance concrete in severe environment. ACI SP 140, Detroit 1993

28. SWAMY, R.N. Alkali-aggregate reaction - the bogeyman of concrete. Proc. V.Mohan Malhotra Symposium in Detroit, ACI SP 144, 1994, 105-131

29. IDORN, G.M. Concrete progress. Thomas Telford, London 1997, 359pp

30. BOURNAZEL, J.P., MORANVILLE, M. Durability of concrete: the crossroad between chemistry and mechanics. Cem.&Concr.Res., 27, 10, 1997, 1543-1552

31. BRANDT, A.M., ed. Optimization Methods for Material Design of Cement-Based composites. Routledge/E & FN Spon, 1998, 314pp

32. NEVILLE, A.M., AÏTCIN, P.-C. High performance concrete - An overview. Materials and Structures RILEM, 31, 1998, 111-117

BEHAVIOUR OF REINFORCED CONCRETE FLEXURAL MEMBERS STRENGTHENED BY BONDED STEEL PLATES AND GLASS FIBRE PLATES

B Jonaitis

V Papinigis

Z Kamaitis

Vilnius Gediminas Technical University

Lithuania

ABSTRACT. The paper examines behaviour of RC flexural members strengthened by epoxy - bonded steel plates or fibre sheets under short-term static loading. Based on experimental and theoretical studies, the model has been proposed to predict the structural response with respect to cracking resistance and strength of flexural strengthened members. The model is based on the behaviour of component structure composed of concrete, internal and external reinforcement as well as the polymer layer between concrete and steel plate. Plate bonding technique has been used in Lithuania to repair many structures in the field. The strengthening of precast prestressed concrete roof slabs 12 m lang of school building is described.

Keywords: Strengthening, Epoxy-bonded steel plates, Glass fibre plates, Cracking resistance, Stiffness, Strength, Roof slabs.

Dr Bronius Jonaitis is an Assoc. Prof. in Departament of Concrete and Masonry Structures of Vilnius Gediminas Technical University, Lithuania. Field of activity: analysis of structures, state assessment, renovation.

Dr Vytautas Papinigis is an Assoc. Prof. in Departament of Concrete and Masonry Structures of Vilnius Gediminas Technical University, Lithuania. Field of activity: analysis of structures, state assessment, renovation.

Professor Zenonas Kamaitis is Director of International Studies Center, Vilnius Gediminas Technical University, Lithuania. Spheres of scientific interest - research of RC bridges and structures, durability, renovation, strengthening and corrosion prevention.

INTRODUCTION

At present, there widespread deterioration of reinforced concrete structures in Lithuania, which has resulted from a large application of reinforced concrete starting from 1950 - 60 s. Many of these structures need to be strengthened.

One of the most interesting and recently developed methods of strengthening existing concrete structures is the externally bonding of additional steel plates or fibre composite sheets with epoxy resin adhesives. [1, 3].

This article deals with influence of external glued reinforcement on crack resistance and strength of normal sections of flexural members - beams. An example of the use steel plate bonding technique to improve stiffness, cracking resistance and load bearing capacity of damaged prestressed roof slabs is described.

EXPERIMENTAL DETAILS

Four series of beams have been tested under short time static load (Figure 1). A series consisted of 2 beams. Beams tested were of 70×100 mm in cross - section and length of the beams was l=970 mm. The beams were made of normal weight concrete of cube strength f_c=41.5 MPa, reinforced with 2∅5 Bp I (reinforcement ratio μ_s=0.00696).

In addition beams were reinforced with external glued reinforcement. The external reinforcement of beams of series S3 were lapped over external ribs of the beams.

Moment of crack formation $M_{cr,obs}$, failure moment $M_{u,obs}$, distance between normal cracks and limit deformations of concrete in tension zone were determined in beam tests.

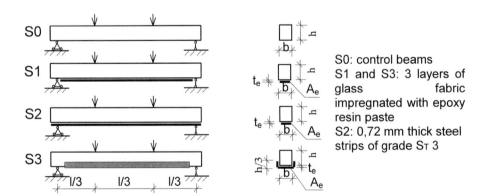

S0: control beams
S1 and S3: 3 layers of glass fabric impregnated with epoxy resin paste
S2: 0,72 mm thick steel strips of grade Sт 3

Figure 1 Test beams with external glued reinforcement

Table 1 The main characteristics of samples and test results

SERIES	CUBE STRENGTH OF CONCRETE	EXTERNAL REINFORCEMENT				TEST RESULTS			
		Type	Strength f_{te}, MPa	A_e, cm^2	$\varepsilon_{e,u}\times \times 10^5$	$M_{cr,obs}$ kNm	$\dfrac{M_{cr,obs,i}}{M_{cr,obs,o}}$	$M_{u,obs}$ kNm	$\dfrac{M_{u,obs,i}}{M_{u,obs,o}}$
S0	41.5	-	-	-	-	0.39	1	2.05	1
S1	41.5	glass fabric	55.6	2.775 2.625	1013	0.78	2	2.77	1.35
S2	41.5	steel strip	f_y=287.7	0.504	400	1.3	3.33	2.21	1.08
S3	41.5	glass fabric		76.48	3.85 2.56	963	0.78	2	2.78	1.35

According to the test results (see Table 1) resistance to normal crack formation of beams strengthened with external glass fabric and steel glued reinforcement increased by 2 and 3.33 times, respectively, in comparison with control beams. Many fine cracks appeared in beams strengthened with external reinforcement. Their width w=0.025 - 0.05 mm and average spacing s_r=55 - 60 mm. Crack width in non - strengthened (control) beams was larger w=0.15 mm and spacing was s_r=85 mm.

Ultimate strength increased 1.35 and 1.8 times respectively. Lower increase in strength of beams strengthened with steel external reinforcement is due to the fact, that the reinforcement at support regions broke off and the bond between reinforcement and concrete surface was damaged.

THEORETICAL CONSIDERATIONS

Resistance of crack formation of flexural members with glued external reinforcement may increase due to stress redistribution between strengthening structure and external reinforcement and also due to increased extensibility of concrete layers saturated with glue. Thus, the cracking moment may be expressed as follows:

$$M_{cr} = M_{cr,c} + M_{cr,p} + M_{cr,s} \tag{1}$$

where $M_{cr,c}$ is the cracking moment of a member without external reinforcement; $M_{cr,p}$ is the moment due to glue effect on the tension strength of concrete; $M_{cr,s}$ is the moment carried by external reinforcement.

Crack formation moment in reference [1] is recommended to calculate as follows:

$$M_{cr} = \frac{W_{pl} f_{br} k_p}{\left[1 - \dfrac{k_z z(z+r)}{\gamma E_c I_{red}}\right]} \tag{2}$$

Where: k_p is a coefficient to allow for tensile strength of concrete covered by polymer; $k_z \leq 1$ is a coefficient to account for deformability (yelding) of the links; z is distance between external reinforcement and the centroid of the cross - section; r is a distance between the centroid of the cross - section and the upper kern point.

For calculation of ultimate strength of flexural reinforced concrete members strengthened by glued external reinforcement we propose to employ the method presented in [2], according to formula

$$M_u = A_e f_{te}\left(h - k_2 x\right) + A_s f_t\left(d - k_2 x\right) \qquad (3)$$

were A_e and A_s are cross - sectional areas of external and internal reinforcements respectively; f_{te} and f_t are strengths of external and internal reinforcements; h and d_0 are the total and effective depths of cross - section; x is a compression zone depth; k_2 is a compression zone parameter. The results of this calculation are presented in Table 2.

Table 2 Results of calculation of crack formation moments M_{cr} and ultimate strength M_u

SAMPLE CODE	CRACK FORMATION MOMENT		STRENGTH	
	$M_{cr,cal}$ kNm	$\dfrac{M_{cr,obs}}{M_{cr,cal}}$	$M_{u,obs}$ kNm	$\dfrac{M_{u,obs}}{M_{u,cal}}$
S1	0.832	0.939	2.856	0.969
S2	0.977	1.33	2.758	0.802
S3	0.858	0.909	3.587	0.786

Typical moment - deflection relationships are shown in Figure 2.   The stiffness of reinforced beams increased 56 per cent for steel plated beams and between 27 and 38 per cent for beams with fibre sheets .

APPLICATION

The use of externally bonded steel plates and fibre sheets as a means to repair damaged reinforced concrete structures was first started in Lithuania in 1982. The first practical experiences in strengthening real structures included bridges and viaducts [1].

The results of our investigations were also used to field application conducted on hall building roof slabs damaged by accident during construction. The vocation school centre which had to be open included one - storey hall building. The main elements of the roof structure of the building are precast prestressed ribbed slabs 12.0 m long, 3.0 m wide, and 0.45 m high (Figure 2, a). The hall building was near completion. Roof accident damage has resulted from the impact caused by falling down of pile of bricks lifting by the tower crane.

Three slabs were significantly cracked. The rupture of steel bars had occurred in places (Figure 2, b).

a)

b)

c)

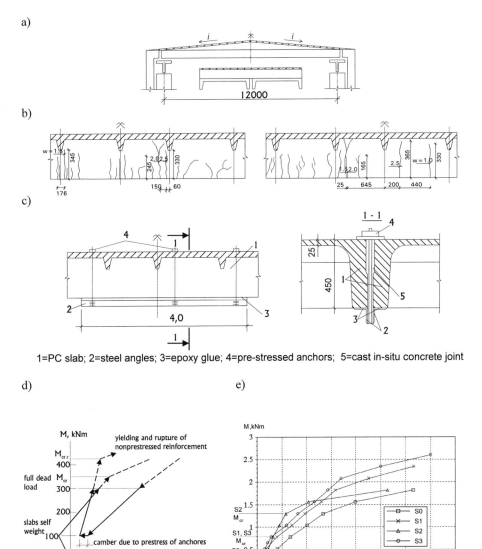

1=PC slab; 2=steel angles; 3=epoxy glue; 4=pre-stressed anchors; 5=cast in-situ concrete joint

d)   e)

Figure 2 Strengthening of PC precast roof slabs: a) cross-section of building;
b) location of cracks; c) addition steel angles; d) moment-deflection plot; e) deflection curves
for unplated (S_o) and plated (S_1, S_2, S_3) beams

Various options were considered. Replacement with new slabs would have been a time consuming, technically difficult, and expensive solution. The steel plate technique was selected. The roof slabs strengthened in pairs with steel angles as tensile reinforcement using epoxy resins (Figure 2, c). The use of steel angles was chosen because of the significant roof slabs deflections and the need of stiffness and load - carrying capacity upgrading. Special attention was paid to the behaviour of slabs near the ends of steel angles to overcome the problem of angles separation. The prestressed anchor bolts were provided at each end of the angles to resist the peeling forces.

The behaviour of the slabs is indicated by load - deflection diagram for mid - span (Figure 2, d). First the slabs were reloaded, cracks were sealed, and external reinforcement was bonded. Then the slabs were loaded by permanent dead load.

The improved stiffness and crack resistance of slabs confirms the effectiveness of the strengthening with external reinforcement.

CONCLUSIONS

Investigation results suggest that external reinforcement is very effective in strengthening of structures. The flexural cracking and strength of test beams increased 2-3 and 1.35 times respectively. The cracking moment and ultimate strength of flexural members can be predicted by equations (2) and (3). External reinforcement was used for strengthening of damaged prestressed concrete roof slabs to return them to nearly the original cracking resistance, stiffness and strength.

REFERENCES

1. KAMAITIS, Z. Repair and strengthening of structures and buildings with syntetic resins. - Vilnius. Technika, 1992, 280 p. (in Russian).

2. DEYRING, M. Versterken von Stahlbeton mit gespanten Faserverbundwerkstoffen. Dubendorf. 1993, p. 163 - 201.

3. McKENNA J K, ERKI, M A. Strengthening of reinforced concrete flexural members using externally applied steel plates and fibre composite sheets - a survey. Canadian Journal of Civil Engineering, vol. 21, No 1, 1994, P. 16 - 23.

EXPERIMENTS ON BOLTED SIDE-PLATED RC BEAMS

M A Bradford
University of New South Wales
D J Oehlers
University of Adelaide
Australia
S T Smith
Hong Kong Polytechnic University
Hong Kong

ABSTRACT. Much of the reinforced concrete (RC) infrastructure is aging. It may be subjected to degradation by harsh environmental factors, or may be required to carry larger loads. It may also be constructed in regions of seismic activity. All of these factors have resulted in industry requiring a means of retrofitting RC beams, and one way by which this is achieved is through steel plating of the sides and soffit. The plates may be glued, or attached by bolts. Attachment by bolts to the sides of beams introduces the unique concepts of vertical (or transverse) partial interaction and partial shear connection, and may also result in buckling of the steel plates between the bolt attachments when subjected to bending, compression and shear.

A series of full-scale experiments has been undertaken at the University of New South Wales with two primary aims: (i) to quantify the increase in strength and ductility that accrues to steel plated construction; and (ii) to study the propensity of the plates to buckle in the presence of bending, compression and shear. The tests have identified a large ductile plateau with a substantial increase in flexural strength, and plate buckles have been observed. These buckles are also unique, as they may only form away from the RC core, and this is referred to in the mathematical literature as a contact problem.

Keywords: Contact problem, Ductility, Plate stability, Reinforced concrete beam, Retrofit, Steel.

Professor Mark A Bradford is Professor of Civil Engineering in the School of Civil and Environmental Engineering at the University of New South Wales, Australia. His main research interests include the stability of steel and composite structures.

Dr Scott T Smith is a Research Fellow in the Department of Civil and Structural Engineering at the Hong Kong Polytechnic University. His recent doctoral dissertation was on the local buckling of steel side plates used to retrofit existing reinforced concrete beams.

Dr Deric J Oehlers is a Senior Lecturer in the School of Civil and Environmental Engineering at the University of Adelaide, Australia. He specialises in fatigue and mechanical shear connection problems in composite structures.

INTRODUCTION

Reinforced concrete beams can be upgraded and repaired by attaching steel plates to their sides and soffit. This technique not only increases the ultimate flexural and shear capacity but reduces deflections due to an increased plate stiffness. Glued and bolted tension or soffit face plating of reinforced concrete beams, which increases the flexural strength, has now become commonly used in industry [1]. The procedure has also been used to repair buildings and strengthen bridges, with use in many countries throughout the world [2]. Only recently have glued side plates, and more recently bolted side plates, been utilised to increase not only the flexural strength of the resulting composite beam, but the shear strength as well.

Upon the initial upgrade of a reinforced concrete beam it is envisaged that plates will be glued and then bolted to the sides and/or soffit of the beam. In the advent of breakdown of the glue, be that from extreme environmental conditions such as elevated temperatures or corrosion of the plate, the bolts will solely provide shear connection. As a lower bound to the analysis of the resulting composite beam it is assumed that only the bolts are effective in providing shear connection. The following study is thus concerned with plates bolted to reinforced concrete beams.

It is within only the last few years that published literature has been available on reinforced concrete beams with bolted side plates, due to its relative infancy in the research fraternity. This is in contrast to the widespread research on reinforced concrete beams with glued tension face plates [1,2,3]. For glued plates full interaction, or zero slip, exists between the steel plate and concrete beam and so design is based on traditional ultimate strength calculations. For bolted plates, partial interaction exists between the steel and concrete interface due to the flexible nature of the bolts. This flexibility provides slip between the two elements and this in turn creates different incompatible strain distributions in the steel and concrete. The major difference between plated beams with a glue connection and plated beams with a bolted connection lies in their different composite action owing to differences in the shear stiffness of the connecting material properties. The behaviour and hence the theory of standard composite beams with mechanical shear connectors [4] and composite profiled beams with rib shear connectors [5], is applicable because their behaviour depends on interfacial slip.

The use of deep side plates, where the plates are subject to in-plane tension, compression and shear can substantially increase both the shear strength and flexural capacity without loss of ductility. Glued side plates suffer from a sudden, almost brittle, mode of failure through concrete failure. Bolted plates on the other hand induce a gradual more ductile mode of failure [6]. Alternatively, shallow plates are predominantly subjected to in-plane tension forces, and although they substantially increase the flexural strength of the composite beam, they reduce the ductility by over-reinforcing the section [7]. As the depth of plate decreases relative to the bottom of the concrete beam, the strength gradually reduces and the mode of failure tends to that of a brittle nature due to concrete crushing. An important criterion in the design of reinforced concrete beams strengthened with externally attached side plates is the plate stability. When the composite beam is subjected to its intended design actions, in-plane compression, bending and shear forces enter the plate through the mechanical shear connectors, and this can lead to instability through local buckling. It is thus essential to position shear connectors at a suitable distance apart to ensure local buckling will not occur until the design strength of the resulting composite plated beam has been reached.

This paper describes a set of tests which serves to quantify the increase in strength and ductility that exists with steel plated construction, as well as to observe the propensity of the plate to buckle in the presence of in-plane stresses. The results of these tests will be presented as well as a comparison to theoretical predictions. This work is of relevance to engineering practice, particularly in Europe and the United States, and will give engineers a feel for the issues that he or she must confront in designing steel side plates for retrofitting RC beams.

EXPERIMENTAL DETAILS

General

The experimental program consisted of 5mm mild steel plates of Grade 250 MPa which were bolted to the sides of full-scale reinforced concrete beams. The beams were 200mm wide × 380mm deep containing 2-Y12 compressive bars and 2-Y24 tensile bars of Grade 400 MPa, and the concrete had a characteristic cylinder strength of 37 MPa. The aim of the experiment was to load the concrete beam several times and observe the local plate buckling. Different regions of plate along the beam were isolated for each loading run with each region being subjected to a unique combination of in-plane axial, bending and shear stress. To ensure local buckling would not occur in an adjacent bay, clamps were applied to the plate to prevent any possible instability. The beams were eventually tested to failure and the load versus deflection characteristics were monitored.

The mechanical shear connectors transferred the load from the concrete beam to the steel plate. The connectors adopted were 90 mm internally threaded adhesive anchors with 10 mm set screws. Since the experiment required progressively removing and inserting set screws, depending on the region of the beam being tested, internally threaded anchors were found to be the most suitable for this purpose. The bolts were snug tightened enough to ensure that full interaction [4] existed between the structural elements, namely that the strain distribution in the concrete was the same as that in the steel. Interfacial slip was thus assumed to be non-existent. The internally threaded anchor bolt arrangement used in this experiment was adopted due to the capacity of the set-screw to be inserted multiple times without damaging the concrete around the hole. The chosen bolt has also exhibited a substantial degree of strength and ductility in experiments undertaken by Ahmed [8] and as such were deemed appropriate for the current experiment.

The current programme involved the testing of two side-plated reinforced concrete beams, whose configuration is shown in Figure 1. The beams were simply supported and subjected to a two point loading arrangement. Three main regions were tested along the length of the beams and they are identified as follows.

- Constant moment or midspan region, denoted as B. The plate here was subjected to combined axial and bending stresses. The adopted two-point loading arrangement produced a constant moment region in the centre of the beam and as such the shear force in the plate equalled zero.

- Shear and high bending region, denoted as SB1. This position was beneath the loading points. The plate was subjected to combined axial, bending and shear stresses.

- Shear and low bending region, denoted as SB2. This position was in the middle of the shear span. The plate was subjected to combined axial, bending and shear stresses.

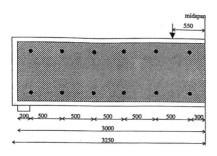 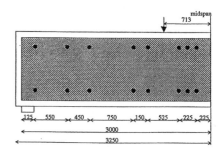

Figure 1 Bolting Configurations for Beams B1 and B2

Set-Up and Instrumentation

Figure 2 shows the side plated beam with full instrumentation just prior to testing. Two linear varying displacement transducers (LVDT's) were attached to the steel plate either side of the beam to measure lateral (out-of-plane) buckling deflections of the steel plate. These LVDT's were moved to the critical buckle region for each new loading run.

One vertical LVDT was placed at midspan to measure the vertical beam deflection and one LVDT to measure vertical slip between the steel plate and reinforced concrete beam near one of the loading points. A horizontal LVDT was also placed at the end of the plate to measure the horizontal slip. These latter LVDT's were used to monitor the amount of slip between the steel and concrete and to verify the full interaction assumption. Full interaction was synonymous with zero slip.

Figure 2 Side-Plated Reinforced Concrete Beam: Experimental Set-Up

Eight electric resistance strain gauges were positioned either side of the beam and down the depth of the plate to measure longitudinal plate strains. The strain gauges were positioned at

Table 1 Specimen details

BEAM	LOAD CONDITION	SPECIMEN DESIGNATION	a (mm)	γ
1	B	BM1-B-a	600	2.0
	SB1	BM1-SB1-a	1000	3.33
	SB2	BM1-SB2-a	1000	3.33
2	B	BM2-B-a	450	1.5
		BM2-B-b	750	2.5
		BM2-B-c	900	3.0
	SB1	BM2-SB1-a	525	1.75
		BM2-SB1-b	600	2.0
		BM2-SB1-c	675	2.25
		BM2-SB1-d	750	2.5
		BM2-SB1-e	750	2.5
	SB2	BM2-SB2-a	750	2.5
		BM2-SB2-b	900	3.0

midspan and under one of the loading points within the critical buckling region. The strain gauges were used to measure the strain distribution down the depth of the beam in the critical buckling region. DEMEC targets were placed down the depth of a section within the middle of the shear span. Their purpose was to measure the strain distribution down the depth of the plate.

One load cell was used to measure the load applied to the beam. This was placed directly under the loading cell. The LVDT, resistance strain gauges and load cell readings were retrieved through a data acquisition system which in turn was connected to a personal computer.

Figure 2 shows the loaded portion of the beam prior to testing. The beam is subjected to a two point loading arrangement approximately at quarter points. The load is distributed from the loading cell by a stiffened universal I-beam to the concrete beam. The concrete beam is thus loaded directly and the load enters the steel plates via the mechanical connectors.

Table 1 gives the specimen designation for each of the two beams tested which is dependent on the applied load. The plate length between the bolts is given in this table as a as well as the aspect ratio γ which is the ratio of plate depth to length between the vertical bolt groupings. Beam 1 is denoted as BM1 while Beam 2 is denoted as BM2.

Test Procedure

Multiple loading runs were undertaken on each beam. Within each run, a particular region of plate was instrumented and allowed to buckle. In order to ensure that buckling would not occur elsewhere in more highly stressed regions along the beam, clamps were applied throughout. For each new loading cycle a different region of plate was isolated and the clamps repositioned accordingly. Several different plate buckling regions were able to be formed by the removal or addition of particular bolts.

All beams were simply supported, 6000 mm in span, and tested in flexure using a two point loading arrangement that produced a constant moment region at midspan. Both beams were designed to fail in flexure and for the plates to buckle well before the ultimate strength of the beam was reached. This was an important consideration in the design of these beams since multiple loading runs were carried out on each beam. The beams were eventually tested to failure and their load versus deflection characteristics were monitored.

TEST RESULTS

Two main test results are reported here and they are the local buckling characteristics of side plated beams as well as their ultimate strength and ductility behaviour.

Plate Buckling

The point of bifurcation, which represents the onset of local buckling was determined by observing the trends of the lateral deflection versus load and in-plane strains versus load curves. Upon bifurcation, the lateral deflection and compressive strain jumped markedly. Here the buckling load and strains were easily determined and the load point predicted with little error.

Following the determination of the local buckling stress at the top and bottom of the plate section, it was then required to express these stresses as equivalent components of axial (compression or tension), bending and shear forces so they could be compared with the theoretical model developed by the authors [9]. The in-plane axial and bending strains were determined directly from the strain profile at the critical sections from the experimental results. At this section, the strain distribution was resolved into two components, namely axial and pure bending. The bending moment was determined about the plate centroid or the mid-depth of the beam and the axial force determined as the difference in the resultant compression and tension force. All the axial resultants were tensile, that is the plate was subjected to a combination of tensile axial stress and bending. The neutral axis in the plate was found to lie above the plate's centroid. The shear force in the plate was determined by calculating the change in bending moment over a length of the plate as given in the following equation. Here two different plate strain distributions were used for this calculation, one within the critical section and the other at the middle of the shear span region.

$$V = \frac{M_{pl} - M_l}{L_l}$$

where V is the plate shear, M_{pl} is the moment in the plate about the centroid in the critical buckling location, M_l the moment in the middle of the shear span and L_l the distance between M_{pl} and M_l. The corresponding buckling stresses were related to the buckling loads by dividing by the plate thickness. The forces determined here were considered as bifurcation loads and are given in Table 2 where P_{ex} is the experimental load required to cause local buckling, n-n the neutral axis in the steel plate at bifurcation (measured from the top of the plate), λ the buckling half-wavelength (determined by the length of the region being tested between attachments) and γ is the plate aspect ratio. σ is the associated buckling stress determined from the buckling forces while the subscript a is the axial, b the bending and s the shear stress component. ex refers to experimental results.

Table 2 Plate bifurcation loads and stresses

SPECIMEN DESIGNATION	BEAM	γ	P_{ex} (kN)	n-n (mm)	$\sigma_{b,ex}$ (MPa)	$\sigma_{a,ex}$ (MPa)	$\sigma_{s,ex}$ (MPa)
BM2-B-a	2	1.5	100	123.3	193.7	34.5	0
BM1-B-a	1	2.0	88	122.8	169.3	30.7	0
BM2-B-b	2	2.5	84	125.3	166.3	27.5	0
BM2-B-c	2	3.0	70	128.3	143.1	20.7	0
BM2-SB1-a	2	1.75	95	127.7	189.4	28.2	7.7
BM2-SB1-b	2	2.0	80	127.6	162.9	24.3	4.5
BM2-SB1-c	2	2.25	70	128.3	143.1	20.7	3.9
BM2-SB1-d	2	2.5	75	127.5	152.6	22.8	4.2
BM2-SB1-e	2	2.5	60	130.2	124.3	16.5	3.4
BM1-SB1-a	1	3.33	55	135.1	111.5	11.1	3.4
BM2-SB2-a	2	2.5	95	108.7	108.5	28.1	5.1
BM2-SB2-b	2	3.0	80	117.2	99.4	25.2	5.0
BM1-SB2-a	1	3.33	85	135.1	111.5	16.7	4.8

Ultimate Strength

Both beams were tested to failure to determine the effect of local buckling of the steel plate on the ultimate strength of the beam. Figure 3 shows one of the beams at failure. Here the concrete has failed by crushing and the steel plate has buckled locally. The ultimate failure loads of Beam 1 and Beam 2 were 143 kN (corresponding to a moment of 164 kNm) and 170 kN (200 kNm) respectively. The load versus deflection response of Beam 1 is given in Figure 4 while that of Beam 2 is shown in Figure 5. Both beams failed in a ductile manner, and this can be seen from the long load plateau. An unplated control beam of the same size and properties was also tested, and its ultimate moment was found to be 133 kNm. Details of this test result are omitted for brevity.

Clamps were applied throughout Beam 1 except within a region of 500 mm below one of the loading points. Upon loading, the plate in this region bifurcated at an applied load of 110 kN, well before the ultimate strength of the beam was reached. No further buckling occurred along the beam due to the restraint offered by the clamps.

The ultimate load of the beam was reached when the post-buckling strength of the plate, in the buckled region, was exhausted and all the load was shed to the concrete beam. The moment capacity of the concrete beam alone thus determined the final strength of the section. The beam eventually failed due to local plate buckling, concrete crushing and large flexural cracking.

For Beam 2, clamps were applied throughout the whole beam to ensure the beam could develop its ultimate strength without premature failure due to local plate buckling as was observed in Beam 1. Buckling however did occur, but after the maximum load was reached, and the points at which they occurred are shown in Figure 4. Upon buckling, a drop off in load was observed. Two sections of plate on both sides of the beam bifurcated, near one of the loading points, and this was then followed by conventional ductile failure of the concrete beam through flexural cracking and concrete crushing.

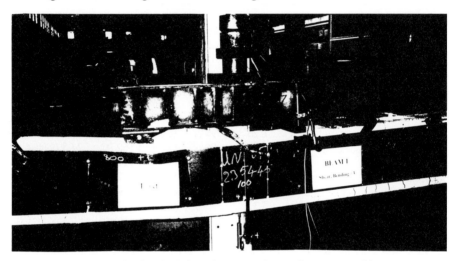

Figure 3 Side plated reinforced concrete beam: Concrete crushing

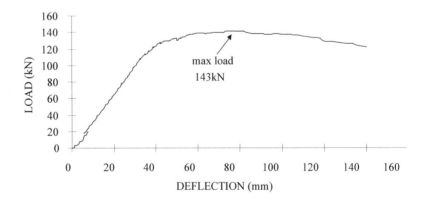

Figure 4 Beam 1: Load versus vertical deflection

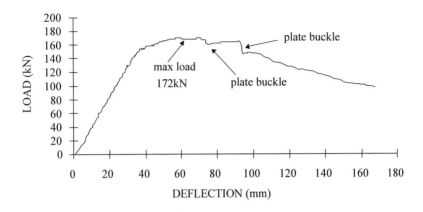

Figure 5 Beam 2: load versus vertical deflection

CALIBRATION OF BUCKLING MODELS

Comparisons between the experimental results and theoretical studies are given in this section. The Rayleigh-Ritz method, as developed elsewhere by the authors [9], was used to analyse the local buckling behaviour of the side plated beams subjected to in-plane axial, bending and shear loading. The analysis gives a dimensionless buckling coefficient to quantify the buckling capacity of the plate at a specific aspect ratio and applied loading. The buckling stress, σ, can be expressed in terms of a dimensionless buckling coefficient, k.

Table 3 presents a comparison of experimental results to theoretical predictions, where k_a, k_b, k_s are dimensionless axial (tension), bending and shear buckling coefficients respectively. The subscripts ex and th denote experimental and theoretical values respectively.

For the purposes of this comparison, the experimental axial and/or shear coefficients were inputed into the theoretical model and the resulting moment buckling coefficient determined which was calculated about the centroid of the plate. Columns 5 and 6 of Table 3 are a direct comparison of the dimensionless moment buckling coefficients while columns 7 and 8 are a comparison of the associated stresses.

The last column in Table 3 is a ratio of the experimental to theoretical bending buckling coefficients and gives an indication of the relative error between theory and test. Each of the buckling stress components are related to their respective dimensionless buckling coefficients by the following equations.

$$k_a = \frac{\sigma_a b^2 t}{\pi^2 D} \qquad k_b = \frac{\sigma_b b^2 t}{\pi^2 D} \qquad k_s = \frac{\sigma_s b^2 t}{\pi^2 D}$$

where D is the flexural rigidity of the plate.

Table 3 Comparison of experimental results and theoretical predictions

SPECIMEN DESIGNATION	γ	$k_{a,ex}$	$k_{s,ex}$	$k_{b,ex}$	$k_{b,th}$	$\sigma_{b,ex}$ (MPa)	$\sigma_{b,th}$ (MPa)	$\sigma_{b,ex}/\sigma_{b,th}$
BM2-B-a	1.5	0.69	0	3.86	5.10	193.7	255.9	**0.76**
BM1-B-a	2.0	0.61	0	3.37	3.77	169.3	189.4	**0.89**
BM2-B-b	2.5	0.55	0	3.31	3.07	166.3	154.0	**1.08**
BM2-B-c	3.0	0.41	0	2.85	2.47	143.1	124.0	**1.15**
BM2-SB1-a	1.75	0.56	0.10	3.77	4.17	189.4	209.4	**0.90**
BM2-SB1-b	2.0	0.48	0.09	3.25	3.55	162.9	178.1	**0.92**
BM2-SB1-c	2.25	0.41	0.08	2.85	3.07	143.1	154.2	**0.93**
BM2-SB1-d	2.5	0.45	0.08	3.04	2.89	152.6	145.0	**1.05**
BM2-SB1-e	2.5	0.33	0.07	2.48	2.66	124.3	133.4	**0.93**
BM1-SB1-a	3.33	0.22	0.07	2.22	1.92	111.5	96.2	**1.16**
BM2-SB2-a	2.5	0.37	0.10	2.16	2.73	108.5	137.2	**0.79**
BM2-SB2-b	3.0	0.34	0.10	1.98	2.31	99.4	115.7	**0.86**
BM1-SB2-a	3.33	0.22	0.10	2.22	1.91	111.5	95.8	**1.16**

The experimental results confirm the general trends of the theoretical findings given Table 3, with the greatest difference in buckling coefficients or stresses being 24%. Most of the other results, however, appear to be within a 10% agreement with the mean of the ratios of test to theory being 0.97 with a standard deviation of 0.14. All plates underwent elastic buckling, that is they buckled before the yield strength was reached. Upon release of the load for each buckling run, the plates returned to their original position indicating no plastic deformation or permanent set. A benefit for this type of plate stability problem is that the buckling capacity of the plate can be increased due to the tension forces produced in this mode of shear connection.

CONCLUSIONS

This paper has reported on a method whereby reinforced concrete beams can be strengthened by attaching steel plates to their sides by bolting. Two full-scale experiments have been described and these demonstrate the increase in strength and ductility that accrues to this type of construction. A failure mode has also been investigated and this is through local buckling of the side steel plates.

ACKNOWLDEGEMENTS

The work described in this paper was supported by a grant made available to the Universities of New South Wales and Adelaide under the Australian Research Council's Large Grants Scheme and a Research Infrastructure Block Grant awarded by the Faculty of Engineering at the University of New South Wales. The second author was also supported by an Australian Postgraduate Award and a Faculty of Engineering Dean's Scholarship.

REFERENCES

1. OEHLERS, D J. "Reinforced concrete beams with plates glued to their soffits", Journal of Structural Engineering, ASCE, Vol. 118, No. 8, 1992, pp 2023-2038.

2. JONES, R, SWAMY, R N AND CHARIF, A. "Plate separation and anchorage of reinforced concrete beams strengthened with epoxy-bonded steel plates", The Structural Engineer, Vol. 66, No. 5, 1988, pp 85-94.

3. HUSSAIN, M, SHARIF, A, BASUNBUL, I A, BALUCH, M H AND AL-SULAIMANI, G J. "Flexural behaviour of precracked reinforced concrete beams strengthened externally by steel plates", ACI Structural Journal, Vol. 92, No. 1, 1995, pp 14-22.

4. OEHLERS, D J AND BRADFORD, M A. Composite Steel and Concrete Structural Members: Fundamental Behaviour, Pergamon, Oxford, UK, 1995.

5. UY, B, AND BRADFORD, M A. "Ductility of profiled composite beams, Part II: Analytical study", Journal of Structural Engineering, ASCE, Vol. 121, No. 5, 1995, pp 876-882.

6. AHMED, M, OEHLERS, D J, AND BRADFORD, M A. "Bolt shear connections for composite plated beam construction", submitted for publication.

7. SMITH, S T AND BRADFORD, M A. "Ductility of reinforced concrete beams with side and soffit plates", Proceedings of the Fourteenth Australasian Conference on the Mechanics of Structures and Materials '95, University of Tasmania, Australia, December, 1995, pp 428-432.

8. AHMED, M. "Strengthening of reinforced concrete beams by bolting steel plates to their sides", Master of Engineering Science Thesis, Department of Civil and Environmental Engineering, The University of Adelaide, Adelaide, Australia, 1996.

9. SMITH, S T, BRADFORD, M A AND OEHLERS, D J. "Local buckling of side plated reinforced concrete beams. I: Theoretical study", Journal of Structural Engineering, ASCE, in press.

FRP REINFORCEMENT - FROM THE LABORATORY TO THE MARKET

J L Clarke

The Concrete Society

P Sheard

Eurocrete Ltd

United Kingdom

ABSTRACT. There is a world-wide interest in the development of fibre reinforced plastic (FRP) materials for the reinforcement and prestressing of concrete, particularly for structures in aggressive environments in which the corrosion of conventional steel can lead to serious problems. This paper reviews the development of FRP materials over the past 10 years or more, with particular emphasis on the EUROCRETE international collaborative research project, and considers the way in which these new materials may gradually be accepted by the Construction Industry.

Keywords: Glass fibres, Carbon fibres, Fibre composites, Corrosion, Durability, Testing, Bridges, Structures.

Professor John L Clarke is a Principal Engineer with The Concrete Society. He has over 25 years experience of the behaviour of reinforced and prestressed concrete structures. His main interest for a number of years has been the development of new materials, including high strength and lightweight concretes and alternative forms of reinforcement. His particular interest has been the development of appropriate design methods for concrete structures reinforced with fibre composites and he has published a number of papers on this and other aspects of the use of FRP.

Dr Peter Sheard is a Director of Eurocrete Ltd. He has pioneered new applications for polymer composite materials for over 15 years and has concentrated since 1992 on non-ferrous reinforcement. His specific technical expertise is in manufacturing processes and corrosion of composite materials in aggressive environments.

INTRODUCTION

Conventional concrete structures are reinforced or prestressed with steel. While it is fully recognised that the steel will corrode in aggressive environments, all design codes work on the basis of assuming that the steel is durable and will not corrode during the life of the structure. This durability is "ensured" in the design process by the requirement to provided adequate protection to the steel in the form of a minimum thickness of concrete of a suitable quality. In many situations this approach fails, leading to costly maintenance and repair of structures. Thus there are incentives to use more durable types of reinforcement, such as epoxy coated, galvanised or stainless steel, though the Construction Industry has been slow to make widespread use of these materials. Over the past 10 or 15 years, there has been considerable interest in the possibility of using fibre reinforced plastic (FRP) materials, which were originally developed for use in the aerospace or automotive industries, in place of steel.

Much of the early work on FRP materials for reinforcement or as prestressing tendons was concerned with the basic properties and the behaviour of simple elements. Some small demonstration structures were built. The materials have now reached the stage that they are going into commercial production. However, the Construction Industry is very conservative, probably with good reason considering some of the problems that have occurred in the past and the long lives that are expected of structures. Thus there are a number of barriers that FRP reinforcement has to be overcome before it is widely accepted by the industry. The paper will cover the key factors, and the steps that are being taken to overcome them; particular reference will be made to EUROCRETE, the pan-European collaborative research project though this will be in the context of developments world-wide.

FIBRE COMPOSITES

Fibres

Currently, the most suitable fibres are glass, carbon or aramid. Each is a family of fibre types and not a particular one. The fibres all have a high ultimate strength (in the region of $3000 N/mm^2$) and a stiffness which ranges from about $70 kN/mm^2$ for glass to over $200 kN/mm^2$ for carbon. The fibres all have a linear elastic response up to ultimate load, with no significant yielding. The fibres are relatively difficult to handle and anchor into concrete and hence are generally combined with resins to form composite (FRP) rods or grids, as described below.

Resins

There is a wide choice of resins available, many of which, though not all, are suitable for forming composites for concrete applications. For example, the draft Canadian Highways Bridge Design Code specifically prohibits the use of polyester resins for embedded FRP material [1]. The choice of resin will depend not only on the required durability but also on the manufacturing process and the cost of the resulting composite. Thermosetting resins are generally used but they have the major disadvantage that, once they have fully cured, the composites can not be bent to form hooks, bends and similar shapes. One possible alternative is the use of suitable thermoplastic resins, which are now being developed.

With these resins the composite components can be warmed and bent into the required shapes. On cooling the full properties of the resins are restored. However, there is likely to be distortion of the fibres in the region of the bend, which will lead to a reduction of the strength locally.

Manufacturing Processes

The most widely used manufacturing process is pultrusion. The fibres, which are supplied in the form of continuous rovings, are drawn off in a carefully controlled pattern through a resin bath, which impregnates the fibre bundle. They are then pulled through a die, which consolidates the fibre-resin combination and forms the required shape. The die is heated which sets and cures the resin allowing the completed composite to be drawn off by suitable reciprocating clamps or a tension device. The process enables a high proportion of fibres to be incorporated in the cross-section and hence a relatively high strength and stiffness is achieved.

As indicated above, thermosetting resins are generally used at present. Once formed these can not be bent into the range of shapes currently used by the concrete industry, and hence different manufacturing processes are required to form specials. Filament winding, in which resin impregnated fibres are wound round a mandrel of the required shape, has been used to manufacture shear links. Other manufacturing processes, such as filament arranging, are being developed for more complex shapes.

Fibre composite two-dimensional reinforcement grids, and even three-dimensional grids, are made by a number of different patented processes.

Short Term Properties

The physical properties of the composite will depend on the type and percentage of fibres used. Typically a pultruded composite reinforcing bar would have about 65% of fibre by volume. Thus with glass the ultimate strength might be $1200N/mm^2$ rising to $1500N/mm^2$ for carbon. The elastic modulus will be about $40kN/mm^2$ for glass fibre composites and may be $130kN/mm^2$ for carbon fibre composites. As composites are not currently manufactured to a common Standard, their properties will vary from one manufacturer to another. Thus all design must be on the basis of the actual properties, as supplied by the appropriate manufacturer.

APPLICATIONS OF FRP IN STRUCTURES WORLDWIDE

Bridges

The widest use of fibre composites in concrete structures has been for prestressing. This is probably because changing from steel to the new materials leads to little, if any, change in the design process. A number of bridges have been built world-wide, generally with conventional steel for the unstressed reinforcement.

In Germany and Austria at least 5 road bridges and footbridges have been built prestressed with glass-fibre composite tendons but with conventional steel for the secondary and shear reinforcement. The first highway bridge, the Ulenbergstrasse Bridge in Dusseldorf, was opened to traffic in 1986 and has been monitored and load tested periodically since [2].

In Japan the emphasis of development has been on carbon or aramid composites. In Japan, at least 10 bridges have been built to date (1998) [3], many with FRP reinforcement as well as prestressing tendons.

The first footbridge in Britain, and probably in Europe, using glass fibre composite reinforcement was built at Chalgrove in Oxfordshire in 1995 [4] as part of the EUROCRETE project. The same project led to the construction of the Oppegard footbridge in 1997 on a golf course near Oslo [5]. It has a span of about 10m with a working load of 5 tonnes and consists of twin arched beams, reinforced with glass FRP rods and stirrups, with horizontal prestressed ties containing aramid tendons.

One of the first significant uses of FRP in North America was in 1993 when part of a bridge in South Dakota, which was stressed with glass and carbon tendons [6]. Later a bridge in Calgary was built with carbon fibre composite prestressing strands [7]. More recently the Taylor Bridge in Manitoba, which has a total length of 165m has carbon FRP for both the prestressing strand and for the unstressed reinforcement in some of the the main beams and the deck [8]. Additional bridges are being built, some of which are considered later [9].

Other Structures

Apart from bridges, fibre composite reinforcement has been used in a number of different applications. The most extensive has been for the reinforcement of sprayed linings for tunnels in Japan, using NEFMAC, which is a grid of glass and/or carbon fibres in resin, which was developed as a direct replacement for conventional welded steel fabric.

In the USA glass fibre composite bars have been used in a number of applications such as chemical plants where high durability is required. In addition it has been used under sensitive electronic equipment where the stray electrical currents in steel reinforcement would be a problem. Descriptions are given in the proceedings of various conferences [10 to 13]

Other experimental structures, which have been built or tested in the laboratory, include post and panel fencing, cladding panels, precast elements for a retaining wall system and a replacement fender support beam for a jetty in the Middle East [5]. A number of other applications are being actively considered.

OVERVIEW OF THE EUROCRETE PROJECT

As indicated earlier, major development work has been carried out around the world. However, much of the work has been fairly fragmented. EUROCRETE, which ran from 1993 to 1997, was the first fully co-ordinated project to bring together all the various aspects concerned with the use of FRP reinforcement in concrete, by bringing together materials suppliers and processors, research organisations, consultants and contractors.

The organisations involved are listed in the Acknowledgements. The key areas of work in the project included:

- The selection of appropriate materials
- The development of appropriate manufacturing processes
- Durability trials, on the raw materials, the composites and reinforced concrete elements
- Structural testing
- The development of appropriate design methods
- The design and construction of demonstration structures

The project involved Partners from the UK, Norway, France and the Netherlands, ran for about 4 years and cost in the region of £4,500,000. Some of the more significant topics covered are described below.

DURABILITY TRIALS

Outline of Philosophy

The requirement of the project was to generate experimental data which could be reliably extrapolated to the likely full service life of a reinforced concrete structure. This could then be used to determine suitable partial safety factors to apply to the short term material properties for use in the design process. To generate the required data, five parallel research programmes were initiated, the overall aim being to obtain a more detailed understanding of the behaviour of embedded FRP rebars with time and hence to develop a realistic model of their behaviour. The five strands of the durability programme were as follows:

- Level 1 - Materials optimisation

- Level 2 - Accelerated rebar durability testing

- Level 3 - Accelerated reinforced concrete testing

- Level 4 - Field exposure testing

- Level 5 - Demonstration structures

The rationale behind the programme was to identify the composite materials most likely to perform well in the concrete environment and to prioritise these initially through a series of accelerated laboratory trials using high alkali fluids. Selected composite materials were embedded in concrete and placed in accelerated exposure conditions to provide an indication of the relationship between the real concrete environment and the selected fluid alkali environment. Levels 4 and 5 were intended to provide data on the behaviour of the selected material in a number of in-service environments. This approach enables the data from accelerated laboratory testing in artificial alkaline environments to be calibrated against the behaviour of the FRP rebars actually in service. Thus the proposed partial safety factors may be adjusted with confidence as more data become available.

Laboratory Testing

The first step was for the resin supplier to determine the most appropriate formulation. Five resin types were assessed in a high pH solution at a range of temperatures.

Over 180 samples were tested over a period of 12 months. A far more extensive programme of testing was initiated on specimens of the composite bar. The variables considered included the effects of stress and the temperature as well as variations in the properties of the fluid in which the specimens were immersed. This is an ongoing programme with a total of about 20,000 specimens in 4 different laboratories.

In addition to the samples immersed in simulated pore water environments, over 1,400 bar samples were cast into concrete specimens, with a range of different concrete mixes and storage environments. The performance of the bars was assessed on the basis of their pull-out strength, it being considered that this would be most indicative of attack.

Field Testing

With any form of durability testing in the laboratory, there is always uncertainty regarding how closely the test conditions represent those in the field. Of particular concern is the approach adopted to accelerate the degradation process. There are simple rule of thumb approaches for determining the effects of elevated temperature on the rate of corrosion, for example, and these can be used in the interpretation of the results. However, to predict the long-term behaviour with any degree of confidence, it is necessary to test specimens in real conditions in parallel with those in the laboratory.

Within EUROCRETE five exposure sites were chosen to provide a range of environmental conditions, both land based and tidal. These are shown in Table 1. At each location a number of samples were exposed, consisting of FRP samples cast into concrete cylinders along with samples containing different types of steel. Half of the samples were returned to the laboratory for testing after about 1 year on site, with the remainder being returned at the end of 2 years.

Table 1 EUROCRETE exposure sites

TYPE OF ENVIRONMENT	LOCATION
Temperate, inland	Roof of industrial building, London, UK
Temperate, intertidal	Estuary of River Severn, UK
Hot, dry	Roof of office building, Dubai, United Arab Emirates
Hot, intertidal	Tidal creek, Sharjah, United Arab Emirates
Arctic, intertidal	Trondheim, Norway

Results from Durability Testing

What follows is a brief summary of the results from the durability testing. Details may be found in Clarke and Sheard [14]. The initial accelerated testing of potential resins in a simulated concrete pore water environment demonstrated a wide range of performances. Some resins maintained a high proportion of their initial properties after one year's exposure while others failed completely. Thus all subsequent testing could be carried out on the "best" resin combined with a suitable fibres.

The testing on the composite bars, both those in the laboratory and those returned from the exposure sites, showed that there was no significant deterioration during the period of testing. In addition, they showed that careful control of the manufacturing process leads to improved, and more consistent, results. Overall they emphasised that the important factor is the durability of the *composite* and not of either the resin or the fibre in isolation.

To date, no samples have been retrieved from any of the EUROCRETE demonstration structures, which are described below. However, on the basis of the results from the laboratory and field specimens it is unlikely that any significant deterioration would have taken place to date. It should be stressed that there have been no reported signs of deterioration on any of the structures: the Chalgrove footbridge was monitored continuously for one year but the only movements were attributed to creep and shrinkage in the concrete and not changes in the reinforcement.

DEMONSTRATION STRUCTURES

While exposure specimens can be used to give a useful indication of the likely in-service behaviour of FRP-reinforced concrete, they are somewhat artificial, having been made under laboratory conditions. There is always some doubt as to how representative "lab-crete" can be of the "real-crete" found in structures, which is significantly more susceptible to the effects of workmanship. Thus the only true method of assessing the long-term performance of FRP in service must be by using the material in demonstration structures, which can be monitored periodically. Within EUROCRETE four demonstration structures were constructed, as shown in Table 2.

Details of the first footbridge may be found in Clarke, Dill and O'Regan [4] and of the third and fourth structures in Grostad et al [5]. Durability samples were attached to the fender support beam which will be removed at intervals for testing.

The Chalgrove footbridge was test loaded to about 1.25 times its design load. Elements of the security fencing (posts and panels) were tested in the laboratory. The data from these tests, along with tests on a large number of beams in bending and shear and other published information, were used to develop the design methods outlined below.

Table 2 EUROCRETE demonstration structures

STRUCTURE	LOCATION	NATURE OF ENVIRONMENT
Footbridge	Chalgrove, Oxford, UK	Inland, temperate, mild winters
Security fencing	Hemel Hempstead, UK	Inland, temperate, mild winters
Footbridge	Oslo, Norway	Inland, temperate, severe winters
Fender support beam	Qatar	Inter-tidal, high ambient temperatures
Non-magnetic foundation*	London, UK	Indoor, buried
Water processing plant*	UK	High humidity, temperate

Note: * indicates that construction took place after the end of the EUROCRETE project

DESIGN METHODS

The current Standards for the design of reinforced concrete structures have developed over the last 100 years or so from simple prescriptive rules to the current mixture of methods based on sound scientific principles and rules of thumb. Some aspects of behaviour, such as shear, are still not well understood and empirical approaches are used. When introducing a new type of reinforcement, with very different properties, one is faced with two alternative approaches; one can adapt the existing approach or one can go back to square one and formulate completely new rules. The latter is obviously the more correct technically, but will be a costly and time-consuming process. Hence it is generally considered that modifying the existing approaches is the only feasible option.

One of the major tasks in the EUROCRETE project was the development of suitable design approaches for FRP-reinforced concrete, modifying the present steel-based rules. The proposed changes have been reviewed and approved by the Institution of Structural Engineers and will be published shortly [15]. Other design codes have been developed in parallel in Japan [16 & 17], in the USA [18] and in Canada [1].

There are areas in which FRP reinforced concrete structures will differ from those using steel. Some examples are as follows:

- Because of the high strength and relatively low stiffness of FRP, it is likely that failure will be by compression of the concrete and not rupture of the reinforcement.
- Bond stresses, and hence anchorage lengths and lap lengths, will depend on the particular bar being used
- Crack widths need be limited only from considerations of aesthetics, and possibly to ensure that the structure is watertight, and not for durability reasons.
- Deflections are likely to be higher than for equivalent steel reinforced units.
- FRP rods have low compressive strengths in comparison to their tensile capacities. Thus the traditional design approaches for columns are no longer valid. Several studies are looking at the effect of FRP wrapped round columns; the confinement leads to significant increases in the failure load and the failure strain.
- Fire will be a significant design consideration for some types of structures. However it is an area that has received very little attention. For embedded reinforcement the prime concern will be to limit the temperature rise at the surface of the bar, so that it stays below the glass transition temperature for the resin.

A further consideration was the choice of appropriate materials partial safety factors, which would reflect the variability of the materials as manufactured and the change in the effective properties with time. On the basis of the durability work described above and other published information, conservative partial safety factors have been proposed. Their use will lead to safe, but non-optimised, structures. As more data become available, both from the laboratory and from experience of the in-service behaviour of actual structures, it will be possible to modify the factors, which will result in more cost-effective use of the materials.

In many situations, reinforcement is only required in the early life of the structure, such as for early thermal crack control. Similarly, in precast units much of the reinforcement may be provided to carry the stresses due to handling, transportation and erection.

In both cases conventional steel can present long-term durability problems and hence the requirements for the amount and quality of the cover are the same as for steel that will be subjected to loads throughout the life of the structure. In addition, the material partial safety factors that are applied during the design are the same in both the long and short term. With FRP it is not reasonable to apply the same factor to reinforcement which is only provided to carry loads early in the life of the structure, before any degradation has taken place. In these circumstances the Institution of Structural Engineers Guide recommends that the factors should be halved.

THE ROUTE TO COMMERCIAL EXPLOITATION

EUROCRETE Limited is now a registered company, which will licence the technology developed during the research project. However, it is envisaged that commercial exploitation will be slow, because of the conservative nature of the construction industry. There is general agreement between those involved with the development of FRP reinforcement that we must move forward slowly, with good quality material used in appropriate applications.

A number of companies have sufficient confidence in the potential of the materials and are now producing FRP reinforcing bars commercially. These include Fibreforce Composites in the UK and Hughes Brothers and Marshall Industries in the USA. In addition, suppliers in the U.K. and elsewhere in Europe are now stocking American bars. A number of Japanese manufacturers produce reinforcing bars, though their main interest has been in prestressing tendons. One such product is currently being used for a 90 m long cable-stayed footbridge being built at Herning in Denmark [19].

One of the areas of the world in which FRP is likely to be more readily accepted is the Middle East, where the high levels of salt combined with the ambient temperature leads to very serious corrosion problems. The possibility of establishing a plant to manufacture EUROCRETE FRP reinforcing bars in the region is currently being investigated.

The construction industry in the UK is extremely conservative, probably with justification when one considers the long time-scales involved. The most likely development route will be the initial use of the new materials in non-structural applications or in ones in which the consequences of failure are not too severe. In North America there is a growing interest in "steel-free" bridge decks, which rely on compressive membrane action to distribute the loads to the main girders. Here the reinforcement is not structural, being only required to control early thermal and shrinkage stresses. This is a very suitable application for FRP reinforcement. An example is the McKinleyville Bridge in West Virginia, which is over 50m long [9].

Another use is for the control of early thermal and shrinkage stresses. In Florida, FRP reinforcement has been in the sprayed-concrete reinstatement of a coastal protection structure.

As confidence in the materials grows, one should be able to move gradually to more highly loaded and critical applications. A likely route may be to introduce the materials initially into precast concrete units, the construction of which is more controlled that that of in situ concrete. However, before FRP reinforcement becomes widely accepted for concrete structures there are a number of significant aspects that have to be addressed, including:

- The ability to produce suitable reinforcement shapes
- The ability to produce large quantities of materials of a consistent quality.
- The development of materials with a standard range of properties
- A better understanding of the behaviour of structures in fire
- The development of design approaches that are more appropriate to the materials

All are essential if the materials are to move forward from their present position as products of a cottage industry to one in which their true potential can be realised.

SUMMARY AND CONCLUSIONS

There is now a considerable experience of the use of FRP materials for prestresssing and for reinforcing concrete structures, with a wide range of demonstration structures built around the world. It is unlikely that FRP reinforcement will ever replace steel for the vast majority of structures in the foreseeable future. However, extensive laboratory testing has shown that, with the appropriate choice of fibres and resins, FRP materials can be durable in the highly alkaline concrete environment. Thus FRP reinforcement is a viable, and cost effective, alternative to steel in special circumstances, for example in highly aggressive environments, where it may be seen as a replacement for epoxy coated or stainless steel bars.

The way forward must be to slowly develop other applications, which can be monitored so that the construction industry can gain the necessary confidence in the new FRP materials. They can then find their rightful place, particularly for concrete structures in aggressive environments.

ACKNOWLEDGEMENTS

Thanks are given to the DTI/EPSRC Link Structural Composites Programme (UK), EUREKA, SENTER (NL), the Ministeere de l'Industrie (F) and the Norwegian Research Council and to the following partners for their generous support and hard work without whom the Eurocrete programme would not have been such a success:

Euro-Projects (LTTC) Ltd.   Laing Technology Group
Sir William Halcrow & Partners Ltd. The University of Sheffield
ASW Construction Systems Ltd.   Techbuild Composites Ltd. (now Fibreforce Composites)
Vetrotex (France)   DSM-BASF Structural Resins
Statoil   Norsk Hydro
SINTEF

Further acknowledgement for additional support is given to:

Norwegian Defence Construction Service   Dr techn Olav Olsen AS
Nordic Supply Composites AS   Norsk Elementbygg AS
Norwegian Public Road Administration   Tarmac Precast
Zoltek   Toray
Geonor

REFERENCES

1. CANADIAN STANDARDS ASSOCIATION, Canadian Highways Bridge Design Code, Section 16, Fibre Reinforced Structures (Final Draft), 1996.

2. WOLFF, R and MEISSELER, H J, Glass fibre prestressing system, Alternative Materials for the Reinforcement and Prestressing of concrete (Ed J.L. Clarke), Blackie Academic & Professional, Glasgow, 1993, pp127-152.

3. TSUJI, Y, KANDA, M and TAMURA, T, Applications of FRP materials to prestressed concrete bridges and other structures in Japan, PCI Journal, July-August 1993, p 50.

4. CLARKE, J L, DILL, M J and O'REGAN, P, Site testing and monitoring of Fidgett Footbridge, in Structural Assessment - The Role of Large and Full-scale Testing, (Eds. K.S. Virdi, F.K. Garas, J.L. Clarke and G.S.T. Armer) E & FN Spon, London,1998, pp 29-35.

5. GROSTAD, T et al, Case studies within Eurocrete - Fender in Qatar and bridge in Norway, (1997) Non-Metallic (FRP) reinforcement for Concrete Structures, Japan Concrete Institute, Sapporo, Japan, 1997, Vol. 1, pp 657-664.

6. IYER, S L, Advanced composite demonstration bridge deck, Fibre reinforced plastic reinforcement for concrete structures (Ed A. Nanni and C.W. Dolan) SP 138, American Concrete Institute, Detroit, 1993, p 831.

7. ANON, Carbon-fibre strands prestress Calgary span, Engineering News Record, 18 October 1993, p 21.

8. RIZKALLA, S. et al, The new generation, Concrete International, June 1998, pp 35-38.

9. THIPPESWAMY, H K, FRANCO, J M and GANGARAO, H V S, FRP reinforcement in bridge deck, Concrete International, June 1998, pp 47-50.

10. NANNI, A and DOLAN, C W, (eds.), Fibre-Reinforced - Plastic Reinforcement for Concrete Structures, Detroit, American Concrete Institute, SP-138, 1993, pp. 977.

11. TAERWE, L, Ed., Non-metallic (FRP) Reinforcement for Concrete Structures, E & F N Spon, London, 1995, pp. 714.

12. EL-BADRY, M M (ed.), Advanced Composite Materials in Bridges and Structures, Canadian Society for Civil Engineering, Montreal, 1996, pp. 1027.

13. JAPAN CONCRETE INSTITUTE, Non-metallic (FRP) Reinforcement for Concrete Structures, Sapporo, Japan, 1997, pp728 and 813 (2 volumes).

14. CLARKE, J L and SHEARD, P, Designing durable FRP reinforced concrete structures, First International Conference on Durability of Composites for Construction, University of Sherbrooke, Canada, August 1988.

15. THE INSTITUTION OF STRUCTURAL ENGINEERS, Interim Guidance on the Design of Reinforced Concrete Structures using Fibre Composite Reinforcement, London, 1999.

16. JAPANESE MINISTRY OF CONSTRUCTION, Guidelines for Structural Design of FRP Reinforced Concrete Building Structures, 1995, (Draft)

17. JAPAN SOCIETY OF CIVIL ENGINEEERS, Recommendations for Design and Construction of Concrete Structures using Continuous Fiber Reinforcing Materials, Concrete Engineering Series 23, Tokyo, 1997.

18. ACI COMMITTEE 440 State -of-the-Art Report on Fiber Reinforced Plastic Reinforcement for Concrete Structures, Report ACI 440R-96, American Concrete Institute, Detroit, 1996.

19. SWIATECKI, S, Building better bridges with CFRP, Reinforced Plastics, March 1998, pp 44-46

ALTERNATIVE FORMS OF REINFORCEMENT FOR BEAMS SUBJECTED TO HIGH SHEAR

K Etebar

Kingston University

United Kingdom

ABSTRACT. In this paper different forms of alternative shear reinforcement other than the conventional shear links have been investigated. These range from a) mild steel plates of varying thicknesses (0.7 mm to 5 mm) with varying numbers of central holes to enhance the uniformity of bond between the steel, concrete and the final product, b) indented mild steel plates for improved bond between steel and concrete, c) profiled plates with shear studs, d) one or more meshes of varying gauges.

Forty eight beams were used in this investigation. Two sizes of test specimens were used, the larger beams having dimensions of 310 x 150 x 2000 mm. The compressive strength of concrete was varied as well as the shear arm ratio to provide an understanding of the effect on the final behaviour. A number of strain measurement techniques were employed on concrete and steel ranging from demec studs to photoelasticity observations to record the changes in the concrete specimens. The results indicate that in addition to increased load bearing capacity, the beams with mild steel plates as shear reinforcement are stiffer, demonstrate less cracking with provision for crack arrest and show adequate bond between plates containing holes and concrete. These are most economical for high shear and not as advantageous for beams being subjected to low shear. The beams with mesh reinforcement also demonstrated better and stiffer resistance to loading.

Observing the limitations of the European and British codes for structural concrete, empirical formula and design calculations are produced to estimate the required size of the mild steel plate as a viable solution to shear links.

Keywords: High shear, Mild steel plates, Alternative forms of shear reinforcement, Shear links, Reinforced concrete beams.

Dr Kamran Etebar is a senior lecturer at School of Civil Engineering, Kingston University. His main research interests include shear in RC beams and structural integrity of repaired concrete elements.

INTRODUCTION

The design of reinforced concrete members against shear failure mainly concerns provision for avoiding a brittle and sudden failure. A reinforced concrete member is able to carry the applied shear through three modes of shear transfer; the dowel action of the longitudinal steel, aggregate interlock across the shear crack and the resistant stresses in the compression zone of concrete [1]. The tension steel provides the stiffness which enables the member to resist the worsening of cracks. This depends on the amount of longitudinal steel and its modulus of elasticity.

Various researchers [2 – 8], particularly within the last two decades have attempted to investigate and explain various concepts of shear behaviour. Primarily emphasis was put on the mode of shear failure, and attention was drawn into the effects of variable design parameters namely:

1. Moment/shear ratio (a/d)
2. Longitudinal steel ratio (A_s/bd)
3. Effective depth (d)
4. Strength of concrete (f_{cu})

In order to evaluate the feasibility of using mild steel plates as shear reinforcement, it is necessary to describe the effect of shear links as a function of shear resistance. Links are beneficial to beam action as they contribute to the strength of the shear mechanisms by the following means:

1. Improving the contribution of the dowel action. A link can effectively support a longitudinal bar that is crossed by the shear crack close to a link.

2. Suppressing flexural tensile stresses in the cantilever blocks by means of the diagonal compression force resulting from truss action.

3. Limiting the opening of the diagonal cracks within the elastic range, thus enhancing and preserving shear transfer by aggregate interlock.

4. Providing confinement to concrete, when the stirrups are closely spaced, thus increasing the compression strength of localities particularly affected by arch action.

5. Preventing the breakdown of bond when splitting cracks develop in anchorages zones because of dowel and anchorage forces.

Why Mild Steel Plates?

Mild steel plates serve as a shear reinforcement for members irrespective of their size and at the same time are protected by a larger cover than the normal links. The plates could assist in reducing the requirement of links and improving the detailing of steel at supports by avoiding congestion of steel. In addition to extra shear capacity these members could benefit from additional ductility which is important for design against accidental loading and for providing an improved residual resistance.

Modes of Shear Failure

Kostovos [9] showed that failure cracks in shear of a concrete beam are associated with multi axial stress conditions that exist in the region of the path along which the compressive force is transmitted from support to support implying that cause of diagonal failure is not entirely dependent on shear force. It appears that shear force is partly contributing to the diagonal failure and the path in which shear forces and compressive forces are transferred from loading point to the support has considerable effect on the diagonal failure.

It is widely accepted that mode of diagonal failures are closely linked with the stresses in the region of concrete cantilevers which are formed between consecutive flexural cracks. Kani [2] in his experiments has shown that mode of diagonal failure is primarily dependent on the shear span/depth ratio and presence of the shear reinforcement. In certain range of a/d ratio (when <2.5, [5]) mode of shear failure may not be dependent on the presence of the shear reinforcement. It only changes the loading value which causes the ultimate failure. The amount of flexural reinforcement and strength of concrete appears to have effects only on shear resistance rather than on mode of shear failure [10].

The extensive shear investigations of Leonhardt and Walter [11] based on measurement of the crack width at various load increments, have shown the beneficial effect of closely spaced stirrups, particularly the diagonal ones. Diagonal stirrups are usually impractical, however, they can provide good resistance against shear. This suggests the provision of mild steel plate can be a close alternative to the use of diagonal stirrups. The ultimate shear resistance of concrete (V_u) is dependent mainly on strength of concrete, on the amount and modulus of elasticity of steel, given in the equation for V_c and the strength provided by the links V_{sv}.

$$V_c = 0.0046(\rho \, E_s \, f_{cu})^{1/3} (400/d \,)^{0.25} \, bd$$

$$V_{sv} = A_{sv} \cdot f_{yv} \cdot d/s$$

b	= width of beam
d	= effective depth
f_{cu}	= characteristic cube strength of concrete
ρ	= $100 A_s/bd$ where A_s = amount of tension steel

It is noticed that unlike other conventional shear reinforcement, steel plates are stressed from the beginning when beam is subjected to loading. Truss mechanism is a valid method which can be adopted to assess the shear resistance of the member [12].

In general, formula can be written as;

$$A_{YP} = k \, (v-v_c) \, b_w d/0.87 f_{YP}$$

where k is an empirical constant which is a function of assumed inclination of the path of concrete strut force and steel tie force. Inclination will depend upon the span, upon the bond and adhesion between plate and concrete as well as the shear span to depth ratio.

EXPERIMENTAL DETAILS

The test programme included 4 sets of beams, each containing 12 specimens, manufactured to investigate the influence of various parameters on the strength of beams containing mild steel or other alternative forms of shear reinforcement. All beams were of rectangular cross sections and to ensure consistency in concrete strength, batches of concrete were mixed to provide adequate quantity for two beams 12 cube 2 modulus of rupture and 3 cylinder specimens. The first series contained 2T16 as tension and 2R6 as compression reinforcement. The beams dimensions were 310 x 150 x 2000 mm. The beams were tested for variable a/d ratios of 1, 2 and 3. There were three beams with no shear reinforcement, three with shear links, three with mild steel plate (2mm thick and 200 mm deep) containing 50 mm diameter holes cut out to improve bond, and three with mild steel plates which where indented using a hammer action.

The second series contained 2R20 as tension and 2R6 as compression reinforcement. Beams were 300 x 150 x 1200 mm in dimension. The effect of various thicknesses of mild steel plates (0.7 - 1.5 mm) and different size of holes (25 and 50 mm diameter) cut out in them was investigated. This series included a beam with nominal links one with no shear links one with maximum allowable shear links and one with mild steel plate with studs welded to the surface for improved bond instead of holes.

The third series contained 2T16 as tension and 2R6 as compression reinforcement. The beams dimensions were 310 x 150 x 2000 mm. The beams were tested for variable a/d ratios of 1, 2 and 3. For each a/d ratio there was a beam with the following shear reinforcement: no shear link, shear links designed to BS 8110, mild steel plate 2mm thick and 200mm deep with 75 mm diameter holes at 200 centres, and 25 x 50 x 3 mm mild steel mesh at full effective depth.

The last series included beams with concrete compressive strength of 30 - 50 N/mm^2. The beam dimensions were 250 x 150 x 1050 mm. Beams with no shear reinforcement, nominal links and shear links to resist a maximum of 5 N/mm^2 were tested as well as ones with mild steel plate and mesh reinforcement. Three beams had single 0.5 - 1.5 mm thick plates with 25 mm diameter holes at 150 centres. The same number had mild steel plates with twice the number of holes and double in the diameter. One beam had a 0.5 mm thick plate with 20 holes of 50 mm in diameter. The last two beams included galvanised wire meshes of 1.62 mm diameter at 25mm centres (one with 2 sheets and one with 4) as shear reinforcement.

Steel moulds were used for construction of all specimens. All beams and cubes were compacted to identical specifications using vibrating tables. Specimens were cured in identical conditions and tested after a 28 day curing period. See Table 1 for beam details.

Instrumentation

One of the important aspect of the testing regime was to obtain measurements for strain and deflection at critical points. In order to obtain strain measurements of concrete, Demec studs were fixed on the surface of the concrete beam, and for strain measurements in steel plate, rectangular rosette Electrical Resistance Strain (ERS) gauges were fixed on the surface. Demec and ERS gauges were fixed to measure both flexural and shear strains at required points and along the critical sections. Two of the beams in series four were instrumented for polariscope readings, where photo-elastic strain patterns show the nature of stress concentrations.

Table 1 Details of the beams used in the investigation

BEAM	% REINFORCE-MENT	A/D	CONCRETE STRENGTH (N/mm²)	1ST CRACK (KN)	ULTIMATE LOAD (KN)	TYPE OF SHEAR REINFORCEMENT
1	1	1	30	170	253	5R8
1A	1	1	30	180	280	MS Plate with holes
2	1	2	32		158	4R6
2A	1	2	32	120	180	MS Plate with holes
3	1	3	33	60	90	10R6
3A	1	3	33	70	117	MS Plate with holes
4	1	1	32	120	250	
4A	1	1	32	155	276	MS Plate :indentations
5	1	2	32		148	
5A	1	2	32	100	180	MS Plate : indentations
6	1	3	30	60	92	
6A	1	3	30	80	115	MS Plate with indentations
B1	1.6	1.5	38	140	239	R6 @ 205 centres
B2	1.6	1.5	38	120	224	
B3	1.6	1.5	35	140-160	272	R8 @ 74 centres
B4	1.6	1.5	38	140	247	MS Plate with 16, 50mm holes
B5	1.6	1.5	38	140-160	232	MS Plate with 16, 50mm holes
B6	1.6	1.5	37	110	221	MS Plate with 16, 50mm holes
B7	1.6	1.5	36	140-160	218	MS Plate with 16, 50mm holes
B8	1.6	1.5	36	140-160	202	MS Plate with 22, 25mm holes
B9	1.6	1.5	37	160-180	248	MS Plate with 22, 25mm holes
B10	1.6	1.5	33	150-180	221	MS Plate with 22, 25mm holes
B11	1.6	1.5	35	150-180	243	MS Plate with 22, 25mm holes
B12	1.6	1.5	36	90-110	218	MS Plate with studs
1A	1.15	1.5	20	130	220	R6 @ 400 mm centres
1B	1.15	2	20	40	130	R6 @ 400 mm centres
1C	1.15	3	27	50	80	R6 @ 400 mm centres
2A	1.15	1	27	50	220	MS Mesh
2B	1.15	2	25	40	150	MS Mesh
2C	1.15	3	25	60	90	MS Mesh
3A	1.15	1	29	50	250	R10 @ 75 mm centres
3B	1.15	2	29	70	130	R10 @ 200 mm centres
3C	1.15	3	27	50	90	R10 @ 200 mm centres
4A	1.15	1	29	175	245	MS Plate with holes
4B	1.15	2	27	60	170	MS Plate with holes
4C	1.15	3	27	50	90	MS Plate with holes
L1	1.9	1.5	38	130	270	R10 @ 65 mm centres
A8	1.9	1.5	48	80	270	MS Plate with 20, 50mm holes
L2	1.9	1.5	52	100-95	265	R6 @ 106 mm centres
A5	1.9	1.5	32	90-80	225	MS Plate with 14, 50mm holes
A3	1.9	1.5	43	90	250	MS Plate with 7, 25mm holes
M10	1.9	1.5	54	80-100	250	MS Mesh
A4	1.9	1.5	31	110-60	195	MS Plate with 7, 25mm holes
A6	1.9	1.5	45	90-110	210	MS Plate with 14, 50mm holes
M9	1.9	1.5	51	80-100	215	MS Mesh
A7	1.9	1.5	43	75-110	185	MS Plate with 7, 50mm holes
0	1.9	1.5	35	95	170	

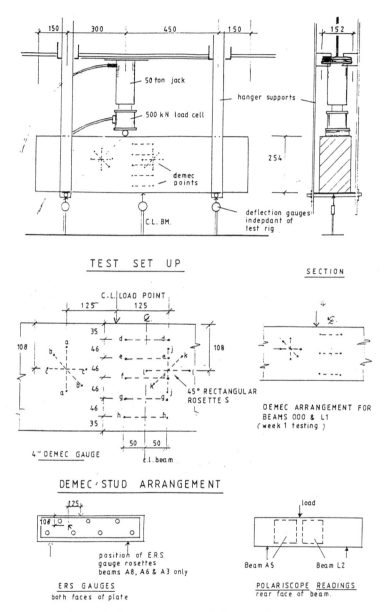

Figure 1   Instrumentation details for the beams

Deflection measurements were taken at the centre span of the beam using a displacement transducer. Figure 1 indicates the position of various strain gauges and the arrangement of testing rig. Details of the plates are shown in Figure 2 . During the test, formation and propagation of flexural and shear cracks were monitored throughout the test. One face of the beam was painted white for easy visibility of cracks. After the tests the beams were carefully broken to investigate the state of bond between concrete and steel plate at failure .

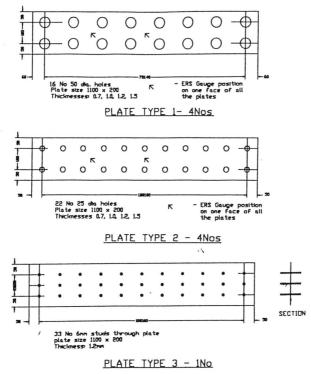

PLATE TYPE 1- 4Nos

PLATE TYPE 2 - 4Nos

PLATE TYPE 3 - 1No

Figure 2   Detail of the plates

Test Results

Trial specimens which acted as reference beams were used in all the four series. There were many observations made from the 48 beams. A commentary on the behaviour of all these beams which were different in some detail would be too vast. Therefore a selection of interesting points as well as common conclusions are described below.

Deflected profiles and load strain plots for beams with various reinforcement are presented in Figure 3. The load strain plots for the main reinforcement are shown in Figure 4. As expected with an increasing a/d ratio the shear capacity of beams in series 1 and 2 decreased. The use of mild steel plates allowed an increase of between 13% for a/d ratio of 1 and 25% for a/d of 3, when compared to a conventionally designed beam with shear links. Results from the first series indicated that the inclusion of plates did not make any difference in the mode of failure of beams with high a/d ratios. Although shear cracks had developed in these beams, they did not propagate the full depth of the beam and flexural failure was the result. A higher load bearing capacity was however recorded. The beams with plate appeared to be stiffer than the conventional ones and this can be seen in Figure 3. It was observed that the beams containing plates with holes developed higher stresses in the steel than the ones with indentations. Giving an indication of the better bond development between concrete and steel as well as concrete acting as a key across the holes. A higher load was required to initiate the first signs of the shear cracks in the beams with plate.

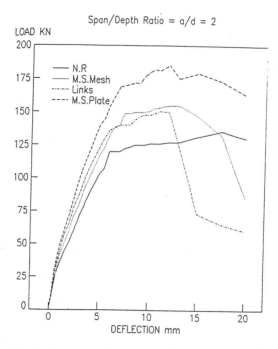

Figure 3   Load deflection curves for beams with various types of reinforcement

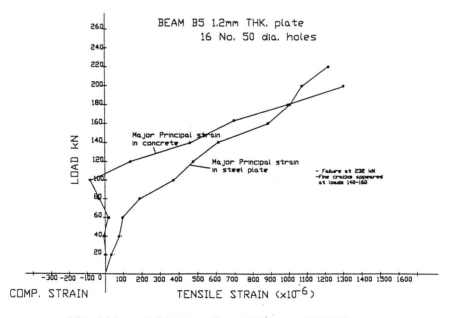

PRINCIPAL STRAIN AT CRITICAL SECTION

Figure 4 Load Strain relationship for plate reinforcement

The disadvantage of using a plate is that in a beam the area of links can be reduced as the shear stress along the beam reduces but this cannot be achieved when using a plate. The same thickness of plate is used throughout the length of the beam.

A similar behaviour was observed in the second series of beams and it is recommended that the use of mild steel plate as an alternative to links be only made in circumstances where high shear is involved, (a/d<2.5). Beyond this a/d ratio, the use of mild steel plates becomes uneconomical, since the percentage of reduction in the weight of the mild steel plate in case of a/d = 1 was 59%, and for a/d = 2 was 2%, and for a/d = 3 there was no reduction in the mild steel plate but an increase in the weight of = 22%.

In the third series it was observed that Sufficient bond requirement needs to be satisfied to allow the use of mild steel plates as shear reinforcement. Perforation in the form of 50mm, and 25mm diameter holes provided sufficient bond between concrete and steel. Larger diameter holes appear to be more efficient than smaller ones. The plates with the studs welded to it did not perform as well as expected . The appearance of the first shear cracks took place earlier than similar beams containing plates with holes. The lower load bearing capacity was accompanied by flexural type cracking, rotation, local bond and bearing failure.

The fourth series demonstrated a few well known facts. The beams reinforced with maximum required shear links showed a slow ductile failure with limited deflection. Reduced surface cracking was observed, although no increase in flexural strength was recorded and the beam used a lot of steel and fixing time. The beams with thinner plates demonstrated better bond through possible indentations as well as the hole provisions. Beams containing meshes had little resistance to spread of cracking due to the yielding of the mesh material. As expected good bond was achieved. It appears more beneficial to try using the meshes near the surface of the concrete as a cage. plates of 1.5 mm thickness provided extra flexural strength (14%) and restrained crack propagation, till the yielding of the plate occurred. The beams with increased rigidity may demonstrate a sudden mode of failure, accompanied with a tearing or yielding of plates and splitting of concrete. One disadvantage of these beams is the lack of confinement of concrete, and hence allowance of the lateral expansion of concrete and weakening of the bond. Separation of the beam into two halves may occur and this would make the height of the plates critical in design. It was also concluded that plates of 1.2 mm and thicker should be employed in these beams.

Detailing Requirements

In order that design satisfies the functional requirements of the structural member, good detailing of reinforcement is essential. Detailing is always based on the thorough understanding of the function of every component in the structural member. Using steel plate as shear reinforcement reduces the confinement of concrete change of shear resisting mechanism was occurring by simultaneous shear mechanism of plate and concrete. Shear links normally are stressed after the concrete has started to crack. With the steel plate concrete should sustain higher load before developing cracks. This is mainly due to the adhesion and bond between a larger surface area of steel plate and concrete with improved shear resistance as a result of development of truss mechanism.

Design Example

Assume a concentrated load of 200 kN, producing max shear on left hand side of;

$$V = 200 \times 525/900 = 117 \text{ kN}$$

Total shear stress $v = 117 \times 1000/ 150 \times 259$
$$= 3.0 \text{ N/mm}^2$$

concrete shear capacity, from BS 8110 Part 1

$$v_c = 0.879 \text{ N/mm}^2$$

enhancement for a/d <2 = 2 d/a = 2 x 259/375 = 1.38
$$v_c = 1.38 \times 0.879 = 1.213 \text{ N/mm}^2$$

shear to be resisted by plate:

$$v_s = v - v_c = 3\text{-}1.213 = 1.789 \text{ N/mm}^2$$

Using formula $A_{sp} = k (v - v_c) bw \; d / 0.87 f_{yp}$

assume k = 1, and $f_{yp} = 275 \text{ N/mm}^2$

$$A_{sp} = 1 \times 1.789 \times 259 \times 150/0.87 \times 275$$
$$= 290 \text{ mm}^2$$

use 225mm deep plate,

thickness of plate required, tp = 290/225 = 1.29mm

Use 1.4 mm thick plate.

CONCLUSIONS

1. Increased load capacity in the beams with plate.

2. Stiffer beams were the result of plate inclusion and hence a lesser deflection recording.

2. Less cracks were developed. Flexural cracks were smaller in width and the initiation of the shear cracks was delayed.

3. Adequate bond between steel and concrete for plates with holes was demonstrated. Although no sign of bond failure was observed for the plates with indentation, the stresses developed in the plate was not as high, giving an indication of higher stresses carried by concrete, and therefore development of more tension cracks.

4. The beams with plates proved more economical for beams with high shear. In beams with low shear there is no real advantage in use of plates.

5. Studded plates and meshes did not perform as well as plates with the correct thickness and will require further investigation before recommendation for use.

REFERENCES

1. ETEBAR, K. The Influence of Aggregate Interlock on the Shear Capacity of Rectangular Reinforced Concrete Beams Without Web Reinforcement. D Phil Thesis, University of Sussex, 1985.

2. KANI, HUGGINS and WITTKOPP, Shear in Reinforced Concrete, Department of Civil Engineering, University of Toronto, 1979.

3. REGAN, P E. Shear in Reinforced Concrete Beams, Construction Industry Research and Information Association, Technical Note 46, London, 1972.

4. TAYLOR, H P J. Investigation of the Forces Carried Across Cracks in Reinforced Concrete Beams in Shear by Interlock of Aggregate, Cement and Concrete Association, 42.447, November 1970.

5. KOSTOVOS, M D. Mechanism of Shear Failure, Magazine of Concrete Research, Vol 35, No. 123: June 1983, pp 99-106.

6. ETEBAR, K. Use of Plates as Shear Reinforcement, South Bank University, Report 1988.

7. ETEBAR, K. Use of Mild Steel Plates for High Shear in Reinforced Concrete Beams, South Bank University, Report 1989.

8. SUBEDI, N K. Reinforced Concrete Beams with Plate Reinforcement for Shear, Proc. Institution of Civil Engineers, Part 2, Vol 87, Sept 1989, pp 377-399.

9. KOSTOVOS, M D, Compressive Force Path Concept: Basis for Reinforced Concrete Ultimate Limit State Design, ACI Structural Journal, Feb 1988, pp 68-75.

10. MPHONDE, A G and FRANTZ, G C. Shear Tests of High and Low Shear Concrete Beams Without Stirrups, ACI Structural Journal, Jul - Aug 1984, pp 350-357.

11. HSU, T T C, The staggering concept of shear design safe, *ACI Structural journal,* Nov-Dec 1982, pp 435-443.

12. EVANS, R H and KONG F K. Shear Design and British Standard Code CP 114, the Structural Engineering Journal, V 45 No 4 April 1967, pp 153-158.

A SOFTENING MODEL FOR FLEXURAL DEFORMATION OF FIBRE REINFORCED CONCRETE

J Y Cui

Taiyuan University of Technology

Y H Zhang

Institute of Water Resources and Hydropower Research

China

ABSTRACT. In this paper, flexural characteristics of fiber reinforced concrete (FRC) are studied. Based on flexural characteristics and ultimate flexural strength of fiber, a softening model predicting flexural deformation curve of fiber reinforced concrete or cement mortar is proposed. Flexural properties of polypropylene fiber reinforced cement mortar is also studied by experiment. With the softening model, flexural toughness of polypropylene reinforced cement mortar are calculated, and they show good agreement with the experimental results.

Keywords: Composite materials, Fiber reinforced cement mortar, Flexural toughness, First crack strength, First crack deformation

Mr Jiang Yu Cui is a Research Fellow in China Civil Engineering Society, Taiyuan, University of Technology, China.His research interests include geotechnical engineering and fiber reinforced concrete. He is currently completing his Ph.D. on the subject of earthquake engineering in China Institute of Water Resources and Hydropower Research, China.

Ms Yan Hong Zhang is a doctor in China Institute of Water Resources and Hydropower Research. Her research fields include structural engineering and earthquake engineering.

INTRODUCTION

The application of fiber reinforced concrete or cement mortar is increasingly more extensive in the field of engineering however studies of properties of fiber reinforced concrete are not yet profound. Nowadays most researchers concentrate their attention on steel fiber reinforced concrete. In some moderated-large cities, the steel fiber reinforced concrete or cement mortar has already been applied in the airstrip, high-speed highway, etc. However, studies of polypropylene fiber reinforced concrete or cement mortar are much poor. The flexural strength and toughness of fiber reinforced concrete or cement mortar are two main performance index. The life-span of the structural member is dominated by the ultimate flexural strength and toughness of the materials in large degree. Therefore, the study of flexural performance of fiber reinforced concrete or cement mortar is important. The characteristic of concrete is brittle failure and its toughness is very low. Fiber reinforced concrete has high first crack strength and toughness. After first crack occurs, the crack-open velocity of the fiber reinforced concrete or cement mortar is much slow. The member of the structure can absorb a large amount of energy at a certain bearing capacity. Especially to the member subject to impact load and vibration load, fiber reinforced concrete or cement mortar has obvious advantages. Fiber reinforced concrete or cement mortar has high flexural toughness. But how to evaluate the toughness of fiber reinforced concrete or cement mortar with a simplest and accurate method play a more important role in the engineering in recent years. Various methods of evaluating the toughness of the composite materials have been proposed. For example, American ACI 544 committee and Japanese JCI society have already put forward different ways to evaluating toughness[1][2]. These methods are mainly used to the steel fiber reinforced concrete or cement mortar. But chemical fiber such as polypropylene fiber (or glass fiber) reinforced concrete (or cement mortar studies) , which are studied much less , has three main advantages: high toughness, high first crack strength and capacity of absorbing a large amount of energy . Therefore, based on the flexural experiment and analyses of polypropylene fiber reinforced cement mortar, a softening model for the flexural deformation of polypropylene fiber reinforced cement mortar has been proposed. With this model, the flexural toughness of fiber reinforced cement mortar at different fiber volume has been evaluated.

EXPERIMENTAL DETAILS

It is known that the curve of flexural deformation of fiber reinforced cement mortar reflects the whole process of slow crack propagation, developing and opening of the composite materials. It is macroscopic behavior of intrinsic structure of fiber reinforced cement mortar in the bearing capacity process, which reflects the ultimate bearing capacity, deformation capacity and toughness of the composite materials. In order to use fiber reinforced cement mortar in the engineering, analysis method of the properties of composite materials must be established. As previously described, one of the most important mechanical properties of fiber reinforced concrete or cement mortar is its high toughness. After first crack, the flexural toughness of the composite materials is dominated by the feature of fiber reinforced concrete or cement mortar. Fiber volume and the features of matrix also affect the parameter. In order to evaluate the flexural toughness of fiber reinforced concrete or cement mortar, a softening model for the flexural deformation of fiber reinforced concrete is proposed in this paper based on the basic characteristics of fiber reinforced concrete and a lot of experimental results .

Figure 1 is a typical curve of flexural deformation of fiber reinforced cement mortar, in which flexural toughness of polypropylene fiber reinforced cement mortar is very high, that is to say, fracture energy determined from the area under the load-deformation curve increase as the fiber volume increases. In figure 1, which the curve can be divided into three stages : (1) linear stage , in which load is in direct proportion to deformation and the first crack load increases as the fiber volume increases; (2) non-linear increase stage, in which the bearing capacities of fiber and matrix have been re-rationed. Before the first crack load, matrix dominates the bearing capacity. After first crack load, the fiber has changed to the dominating action of the bearing capacity. (3) non-linear decrease stage, in which the bearing capacity of the member begins to decrease when it gets to the ultimate load. At this time, all the fibers are in full action. Due to the Visio-elastic properties of fiber, the load decreases very slowly, which shows that flexural toughness of fiber reinforced cement mortar is very good.

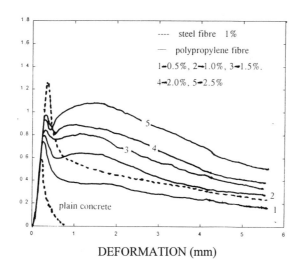

Figure 1 The flexural load-deformation curve of fiber reinforced cement mortar

SOFTENING MODEL OF FLEXURAL DEFORMATION

As analyzed above, the first crack load changes with different fiber volume. The definition of first crack load is different for different researchers, because first crack point is greatly affected by instruments and others factor. According to literature [3], first crack load is defined as follows:

$$P_{cu} = \gamma_{cr} P_{mu} \tag{1}$$

where γ_{cr} is the first crack strength ratio of fiber reinforced cement mortar and matrix can be expressed as:

$$\gamma_{cr} = \sqrt{\frac{V_{f0}}{V_f}} \left[\frac{1 + C_\theta \frac{1}{\sqrt{V_{f0}}}}{1 + C_\theta \frac{1}{V_f}} \right]^2 \qquad (2)$$

where V_{f0} = critical fiber volume;

V_f = fiber volume;

C_θ = fiber distribution coefficient in the matrix, which is 8.85, 11.1 and 12.5 espectively when fiber distribution is one dimension, two dimension and three dimension. For fiber volume $V_f \le V_{f0}$, $\gamma_{cr} = 1$.

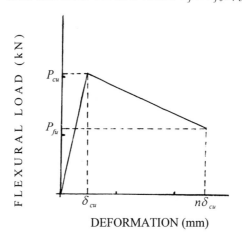

Figure 2 Softening model of fiber reinforced concrete

Based on previous analysis, the softening model equation is as follows:

$$\begin{cases} P = \dfrac{P_{cu}}{\delta_{cu}} \delta & (\delta \le \delta_{cu}) \\[4mm] P = P_{cu} + \left(\dfrac{P_{cu} - P_{fu}}{\delta_{cu} - n\delta_{cu}} \right)(\delta - \delta_{cu}) & (\delta_{cu} \le \delta \le n\delta_{cu}) \end{cases} \qquad (3)$$

where P_{fu} be derived from the beam flexural deformation in the mechanics of materials. as follows:

$$P_{fu} = \alpha V_f \frac{bh^2}{3a} \sigma_{fu} \qquad (4)$$

where δ_{cu} =   first crack deformation;
       $n\delta_{cu}$ =   the deformation of specimens which lose its serviceability;
       P_{fu} =   fiber ultimate bearing capacity at loss of serviceability;
       α =   the percent of fiber ultimate strength when the specimen failures;
       b,h,a =   width, height and distance between load and fulcrum of the specimen;
       σ_{fu} =   fiber tensile ultimate strength.

That the specimen loses its serviceability implies that the member must be dominated not only by its strength but also by its deformation. For example, to polypropylene fiber reinforced cement mortar, although load decreases slowly in a large deformation scale, deformation can not exceed a certain limit so that the member can work .naturally.

EXPERIMENTAL RESULTS

The polypropylene fiber reinforced cement mortar is used in the experiments. The dimensions being $280 \times 50 \times 25mm$, polypropylene fiber length L=25mm, diameter d=20 μm, tensile ultimate strength $\sigma_{fu} = 438\,MPa$, polypropylene fiber is produced by Qi Xian county, ShanXi province. Cement label $325^{\#}$ is produced by Taiyuan, ShanXi province.

The specimens are vibrated on the vibration table. These specimens are cured in water at $20^{0}c$ and subject to four point loading over a pan of 300mm, loads are applied by electronic testing machine with a displacement controlled. The curve areas obtained by experiments are compared with proposed model as showed in Table 1. According to the features of polypropylene fiber reinforced cement mortar flexural load-deformation curve, $P_{mu} = 0.71KN$, $\delta_{mu} = 0.125mm$, $\delta_{cu} = P_{cu}/P_{mu}\,\delta_{mu}$, $n\delta_{cu} = 3mm$, $\alpha = 1/2$. The toughness of the experiment and calculation are showed in table 1.

Table 1. The flexural toughness of polypropylene fiber reinforcement mortar (KN-mm)

V_f (%)	0	0.3	0.5	1	1.5	2.0	2.5
Experimental	45	1360	1500	2150	2750	3450	3900
Caluculated	44	1311	1475	2135	2790	3339	3941

In Table 1, the flexural toughnesses that obtained from model evaluation are in good conformity with the experimented results. Therefore, the proposed softening model in this paper evaluating the flexural toughness of fiber reinforced cement mortar is satisfactory with practice engineering.

CONCLUSIONS

Given the ultimate load, first crack deformation of the matrix and fiber ultimate tensile strength and coefficient α, the formulae suggested in this paper enable the flexural toughness of fiber reinforced cement mortar to be calculated. More experiments are unnecessary for different volume of fiber. The formulae are simple as well as convenient. Provided the features of plain concrete are given, it is possible to calculate the toughness of different fiber reinforced concrete or cement mortar by the proposed formulae. Analyses of other fiber reinforced concrete or cement mortars have not been made and this is an area where further research is required.

REFERENCES

1. ACI COMMITTEE544. State of the Art Report on Fiber Reinforced Concrete. In: ACI Manual of Concrete Practice, Part5, 1985

2. XU JIANHUA Translation. JCI Standard of the Experiment Methods on Fiber Reinforced Concrete. Harbin Institute of Architecture and Engineering 1986, 8

3. CUI JIANGYu GUO WEN WU. The First Crack Strength of Fiber Reinforced Cement Matrix Composite Materials. The Journal of Shi Jia Zhuang Railway Institute. 1994, Vol. 7, No. 3

ANCHORING OF EXTERNALLY BONDED CFRP REINFORCMENT

D Van Gemert

O Ahmed

K Brosens

Catholic University of Leuven

Belgium

ABSTRACT. On using the externally bonded reinforcement for strengthening concrete structure elements, the significant affecting parameter that controls the obtained strengthening efficiency is the anchoring stress. This shear stress is generated at the end region of the longitudinal external reinforcement. Although there is a lot of information available on the behaviour of r.c. structure elements strengthened with externally bonded steel plates, little is known about their behaviour when carbon-fiber-reinforced-plastic (CFRP) laminates are used as a substitute for steel plates. So an experimental investigation was carried out on five r.c. beams and twenty four shear test specimens to study the effect of using this modern materials, CFRP laminates, as well as the different proposed techniques implemented at the end region of longitudinal CFRP laminates to alleviate or overcome the anchoring stress concentration generated at this region on the behaviour of strengthened beams. The paper also includes the discussion and evaluation of the modified formula suggested to predict the load carrying capacity of the beams strengthened with externally bonded CFRP laminates. The correspondence between experiments and predicted values is superior to the predictions by formulae available in literature up to now.

Keywords: Carbon-fiber-reinforced-plastic (CFRP) laminates, Anchorage shear stress, Strengthening, Shear capacity, Flexural capacity, Anchorage capacity, Shear-span.

Professor Dionys Van Gemert is professor of building materials science and renovation of constructions at the Department of Civil Engineering of K.U.Leuven. He is head of the Reyntjens Laboratory for Materials Testing. His research concerns repair and strengthening of constructions, deterioration and protection of building materials, concrete polymer composites.

Eng Omar Ahmed is assistant lecturer at Civil Engineering Department, Assiut University, Egypt. He is currently Ph.D. student at the Department of Civil Engineering of K.U.Leuven. His research concerns strengthening of R.C. beams by means of externally bonded CFRP laminates.

Ir Kris Brosens is currently Research Engineer and Ph.D. candidate at the Department of Civil Engineering, K.U.Leuven. His research concerns repair and strengthening of concrete structures using externally bonded steel plates and carbon fibre reinforced laminates.

INTRODUCTION

CFRP laminates technique can solve many problems encountered in the steel plate bonding technique, e.g. strengthening of r.c. slabs inside box girders; strengthening of two ways slabs [1]; corrosion protection.

The common disadvantage of using CFRP laminates technique is its higher costs in comparison with that of the steel plates technique, but if we try to benefit of the higher strength of the used CFRP laminates with maximum efficiency, this will compensate the difference in costs. This research aims to alleviate or overcome the anchoring stress concentration generated at longitudinal CFRP laminates end region by application of different suggested systems. Tests on a series of r. c. beams with externally bonded CFRP laminates and shear test specimens have been conducted to study the extent of exploitation of using CFRP laminates to enhance the anchoring capacity of the strengthened beams, so enhance the capacity of such beams. The pattern of cracks and mode of failure, the anchorage shear stress distribution, the maximum deformation and the load carrying capacity have been studied. Also a modified equation is proposed to predict the load carrying capacity of the CFRP strengthened beams.

EXPERIMENTAL PROGRAMME

Test Program

Five r.c. beams, designated as AF.0, AF.3, AF.3-1, AF.3* and AF.3** in Figure 1 and Table 1 with a rectangular cross-section of 125 mm width and 225 mm height and total length of 1700 mm, were tested under a four point bending test over a simple span of 1500 mm (shear-span to depth ratio, a/d, equals 2.5).

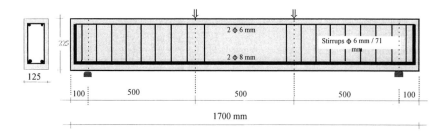

Figure 1 Details of reinforcement of tested beams

These beams were tested to study the effect of the different proposed techniques implemented at the end region of longitudinal CFRP laminates on the anchorage capacity. The first beam, AF.0, was tested in its original condition as a control one. The rest of beams were strengthened with two layers of longitudinal CFRP laminates, each of 0.167 mm thickness (Tonen FTS carbon fiber sheet) and 75 mm width. The lengths of both layers are equal to 1300 mm for beam AF.3. The common anchorage system for beam AF.3 (reference system)is shown in Figure 2. Beam AF.3-1 was provided with two U-shape anchorage side laminates that were bonded at both of the end regions of the longitudinal CFRP laminates (the first proposed system), see Figure 2. The second proposed system (AF.3*) provides a shortening in the outer CFRP laminate layer by 150 mm from both edges, see Figure 3. The last proposed system, AF.3**, is a combination between the previous two systems, AF.3-1 and AF.3*, see Figure 3.

Table 1 Details and data of tested beams

BEAM NO.	DATA OF TESTED BEAMS		f_c, f_{ctm}, E	INTERNAL REINFORCEMENT		LONG. CFRP LAMINATES		
	d (mm)	a/d	(N/mm²)	A$_S$	S(mm)	A$_f$ (mm)	L (mm)	REFERENCE
AF.0	-	-	-	-	-	-	-	Reference 1
AF.3	-	-	48.0	-	-	25.05	100	Figure 2
AF.3-1	200	2.5	2.83	2ϕ8 mm	71	25.05	100	Figure 2
AF.3*	-	-	30000	-	-	25.05	100	Figure 3
AF.3**	-	-	-	-	-	25.05	100	Figure 3

f_c, f_{ctm} and E measured after 42 days from casting.

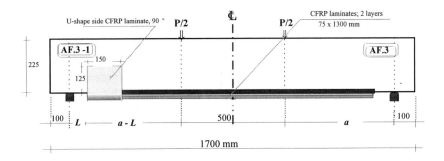

Figure 2 Details and arrangements of bonded-on CFRP laminates
for beams AF.3 and AF.3-1

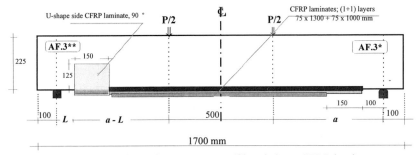

Figure 3 Details and arrangements of bonded -on CFRP laminates
For beams AF-3* and AF.3**

Materials

Ribbed bars (BE 500) of 8 mm and 6 mm in diameter for tension reinforcement and 6 mm for both compression and shear (stirrups) reinforcements. Both yield and ultimate strengths as well as Young's modulus for each diameter are mentioned in Table 2.

The external reinforcement was a CFRP laminate under a commercial name of TONEN carbon fiber sheet, FORCA TOW SHEET (FTS). Such FTS sheet is available in rolled laminate of 0.167 mm effective thickness, 500 mm width and 100 m length. The effective thickness gives the section of the fibers in each ply. The yield strength and Young's modulus of such CFRP laminate are reported in Table 2.

Table 2 Properties of used reinforcements

PROPERTIES	INTERNAL REINFORCEMENT		EXTERNAL REINFORCEMENT (CFRP LAMINATE)
	ϕ 6 mm	ϕ 8 mm	
Yield Strength (N/mm^2)	552.67	568.33	3500
Ultimate Strength (N/mm^2)	613.33	654.00	3500
Young's Modulus (kN/mm^2)	195	185	240

An epoxy mortar (CARFICOM TM, product of De Neef CONCHEM) layer of 2.5 to 3.5 mm thickness was applied to all the strengthened beams as a substratum to the CFRP laminates. This epoxy mortar is completely cured within a period of 24 hours after application. Then the compressive, bending and tensile strengths as well as Young's modulus of this epoxy mortar amount to 80, 20, 6.5, 7200 N/mm^2 respectively. The rounding radius at the lower corners of the U-shaped stirrups was 30 mm minimum.

The resin used for the impregnation of the CFRP laminates was available under a commercial name of CARFICOM RES, product of De Neef CONCHEM too.

ANALYSIS AND DISCUSSION OF TEST RESULTS

Both the influence of the different proposed techniques implemented at the end region of longitudinal CFRP laminates on the behaviour of tested beams and the suggested modified equation to predict the load carrying capacity of the strengthened beams are discussed. Table 3 shows the obtained cracking, growth (end laminates shear crack initiation) and maximum loads (P_{cr}, P_{growth} and $P_{max.}$), cracking and maximum deflection (δ_{cr} and δ_{max}) and the mode of failure for the tested beams. Also, the predicted load carrying capacity (P_{prop}) obtained from the suggested equation as well as the maximum anchorage shear stress (τ_{anch}), shear stress due to shear force (τ_{SF}) and the total shear stress (τ_{tot}) of tested beams are included.

Table 3 Results of tested beams

BEAM NO	P_{cr} kN	P_{growth} kN	$P_{max.}$ kN	P_{prop} kN	τ_{anch} N/mm^2	τ_{SF} N/mm^2	τ_{tot} N/mm^2	δ_{cr} mm	δ_{max} mm	FAILURE MODE
AF.0	22.5	-	55.0	-	-	-	-	1.07	6.61	Flexure
AF.3	25.0	65.0	96.5	93.3	2.09	0.56	2.65	1.08	13.6	Laminates end shear
AF.3-1	30.0	75.0	118	119.0	2.75	0.68	3.43	0.94	20.0	Laminates separation
AF.3*	25.0	80.0	98.5	97.8	2.13	0.57	2.70	1.16	12.0	Laminates separation
AF.3**	25.0	50.0	105.0	119.0	2.35	0.61	2.96	0.79	13.0	Laminates separation

δ_{max} is the mid-span deflection corresponding to the maximum load.

CFRP Laminates Normal Stress and Anchorage Shear Stress

The values of the CFRP laminates normal stresses were derived from the strain measurements. It is worth noting that the end laminates stress increases with the increase of the applied load. The rate of increase continued for both the first and the third systems, AF.3-1 and AF.3**, while such increase in the normal stress continued up to the level of the end laminate shear crack initiation in case of both the reference and the second systems. Such shear crack initiated at a medium level of loading for the reference system, but for that of the second system it initiated at a higher level of loading. At the level of the maximum load, the CFRP laminates normal stress amounted to the compression value for both of the latter systems (AF.3 and AF.3*) but the value of drop in stress in case of the reference system was more than that of the second. The other two systems, providing an anchorage side laminate at the longitudinal CFRP laminates end region, showed a decrease in the rate of increase of the normal stress but still tension up to the level of the maximum load.

The anchorage shear stress between the external CFRP laminates and concrete along the bonded length were derived from both the normal stresses σ and the thickness t_f of CFRP laminates. For the various proposed systems, the anchorage shear stress distribution is plotted along the bonded length for the different levels of loading in Figures 4 and 5.

In fact, shortening of outer laminate proposed system showed an alleviation of shear stress concentration while U-shape anchorage side laminate system provides a considerable overcome of such stresses concentration, see Figures 4 and 5.

At the CFRP laminates end region, the proposed systems showed an anchorage shear stress smaller than that of the reference system up to the level of end laminates shear crack initiation of the reference system. This effect was considerably clear in case of the first proposed system. Also the proposed systems showed a smaller drop in the anchorage shear stress at the laminates end region, just before laminate separation, in comparison with that of the reference system.

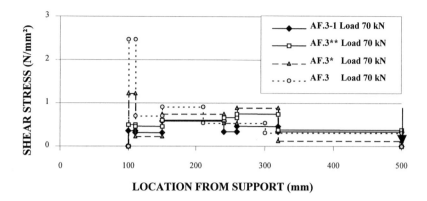

Figure 4 Shear stresses along the bonded CFRP laminates for the different proposed systems

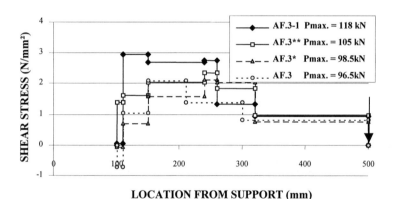

Figure 4 Shear stresses along the bonded CFRP laminates for the different proposed systems at the maximum load

Failure Behaviour

The first crack initiated as a flexural one at the middle third (region of constant moment), in case of both reference and strengthened beams. The mode of failure was flexure for the reference beam, AF.0. However, it was laminates end shear for the reference system, AF.3, see Figure 6. Beam AF.3* failed due to laminates separation. This is attributed to the fact that the critical section, section of higher shear stress, was shifted away from the end region. The failure of beam AF.3** was due to cut off of CFRP laminate at the section adjacent to the inner edge of the anchorage side laminate accompanied with laminates separation. This is attributed to the formation of the main shear crack which started close to the inner side of the anchorage side laminate. This crack results in a relative displacement, v, which leads to cut off of CFRP laminate, see Figure 8. The same behaviour occurred in case of the first proposed system, AF.3-1, but there is no cutting off of the longitudinal CFRP laminates, see

Figure 7. This is attributed to the relative higher resistance for the section at a region close to the inner side of the anchorage side laminate because both of the two CFRP layers were continued to the end of side laminate. The anchorage side laminate was partially cut in case of the first system whereas no clear cut occurred in case of the third system.

The critical section in case of both the first and the third proposed systems was adjacent to the inner side of the anchorage side laminate while it was at the region of the outer laminate end in case of the second proposed system. The failure occurred at a maximum total shear stress of 2.65, 3.43, 2.70 and 2.96 N/mm^2 for the reference and proposed systems respectively. The total shear strength in case of both the reference and the second proposed system, was more or less close to the surface tensile strength of used concrete, 2.83 N/mm^2. However, it was somewhat higher in case of both the first and third proposed systems. This is attributed to the fact that the critical section was adjacent to the anchorage side laminate that strengthened and supported the surface tensile strength in that very short area.

Figure 6 Crack pattern of beam AF.3 (laminate end shear)

Figure 7 Crack pattern of beam AF.3-1 (laminate separation)

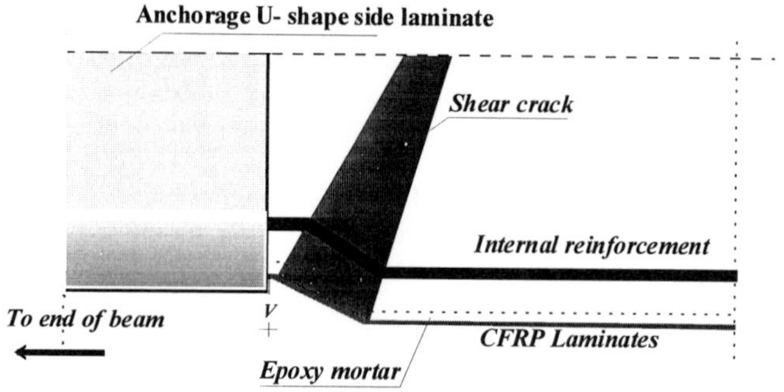

Figure 8 Shear crack relative displacement

The Modified Mathematical Model

The common dominant failure mode of r.c. beams strengthened with externally bonded CFRP laminates is either laminate end shear or laminates separation. So, the significant parameter affecting the behaviour and the capacity of strengthened beam is the shear stress in the adhesive layer between the external reinforcement and the concrete. The critical section locates at the longitudinal laminates end region, where the shear stress concentration occurs. Such shear stress is partially due to the variation of bending moment and the remainder due to the introduction of forces in the anchoring zones (anchorage shear stress). The first portion of shear stress has a considerable role in case of steel plates technique. On the contrary, that portion of shear stress is considerably smaller in case of CFRP laminates. This is attributed to the lower thickness or by other words the lower stiffness of CFRP laminates for the same required degree of strengthening, where the CFRP laminates strength amounts to seven to ten times that of steel plates.

Because of the interpretation mentioned before, a proposed modified equation to predict the capacity of beams strengthened by means of externally bonded CFRP laminates was derived through Eq. (3) suggested by Jansze [2]. The proposed modified expression relies mainly on adding the difference between the portion of shear stress generated due to shear force in case of substituting of CFRP laminates with steel plates of 550 N/mm^2 yield strength (the case of Jansze's study) and that for CFRP laminates, to the shear strength calculated from Equation (3). Therefore, the shear strength τ_{PES} (P = plate) calculated from Equation (3) will be substituted for τ_{FES} (F = fiber). The authors propose to add the modification $\Delta\tau_{MOD}$, given in Equation (5) to obtain τ_{FES} instead of τ_{PES}, see Figure 9 and Equations. (1), to (5).

The calculated theoretical shear span (a_L) according to Equation 4 may be greater than the true shear span as a result of using a small amount of internal tension reinforcement or greater unsheeted length or both. In such case the experiments showed that an imaginary value (a_{LM}) equal to the average of the calculated fictitious shear span and the true shear span, should be applied in Equation 3.

$$F_{FES} = \tau_{FES}\ b\ d_s \tag{1}$$

$$\tau_{FES} = \tau_{PES} + \Delta\tau_{MOD} \tag{2}$$

$$\tau_{PES} = 0.18 \sqrt[3]{3 \frac{d_s}{a_L}} \left(1 + \sqrt{\frac{200}{d_s}}\right) \sqrt[3]{100 \, \rho_s \, f_c} \tag{3}$$

$$a_L = \sqrt[4]{\frac{\left(1 - \sqrt{\rho_s}\right)^2}{\rho_s} d_s L^3} \not> a \tag{4}$$

$$\Delta\tau_{MOD} = \tau_{PES} \, b \, d_s \left(\frac{S_P}{I_P \, W} - \frac{S_F}{I_F \, W_m}\right) \tag{5}$$

where, S_F is the statical moment of area in case of CFRP laminates
I_F is the second moment of inertia in case of CFRP laminates
S_P is the statical moment of area in case of substituting CFRP laminates with steel plates
I_P is the second moment of inertia in case of substituting CFRP laminates with steel plates
t_p is the equivalent steel plate thickness $= t_F (\sigma_F / \sigma_P)$, $\sigma_P = 550$ N/mm^2
f_c is mean compressive cylinder concrete strength
ρ_s is the ratio of continuos tension reinforcement
W, W_m are the widths of CFRP and epoxy mortar respectively (see Figure 9(b))

In case of the beams strengthened by providing anchorage side laminates at the end region of the longitudinal CFRP laminates, the load carrying capacity will be a summation of the load calculated according Eq. (1) and the load due to the anchorage laminate, calculated according to Equation (7). The length of the anchorage side laminate taken into account in the calculation should be not more than the maximum anchorage length calculated from Equation (8).

$$P_{total} = P_{FES} + P_{anch} \not> P_u \tag{6}$$

$$P_{anch} = A_{anch} \, \sigma_{yl} \tag{7}$$

$$l_{max} = \sqrt{\frac{E_F \, t_F}{4 \, f_{ctm}}} \tag{8}$$

in which, A_{anch} is the cross-sectional area of the anchorage side laminates
σ_{yl} is the yield strength of the CFRP laminates
E_F, t_F are the elastic modulus and the thickness of the CFRP laminates
f_{ctm} is the surface tensile strength of concrete in N/ mm^2
P_u is the smaller of either shear capacity or flexure capacity of the
strengthened beam (P = 2 F)

The predicted load carrying capacity of the tested beams (P_{prop}) using the proposed modified equation (1) were calculated and reported in Table 3. It is worth mentioning that the calculated results according to the proposed modified expression achieved a considerable approach to the true values (experimental results), particularly in case of beams strengthened with longitudinal laminates only. However the calculated results were not accurate enough in case of beams provided with anchorage side laminate. This is attributed to the fact that the failure in such beams was due to the relative displacement resulting from the shear crack formed at anchorage laminate inner edge instead of that formed at the laminates end which leads to laminate end shear or laminates separation failure.

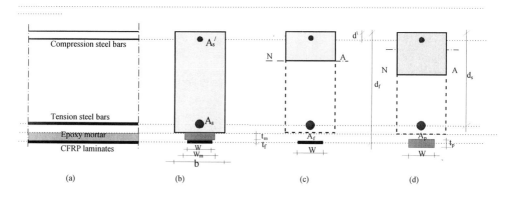

Figure 9 Details of strengthened beam, (a) longitudinal section, (b) cross-section, (c) cracked cross-section, (d) assumed equivalent cracked cross-section

FRACTURE MECHANICAL APPROACH

The anchoring capacity of externally bonded CFRP laminates can also be determined using non-linear fracture mechanics. In [3] a design model for bonded CFRP-concrete connections subjected to pure shear was presented. This model was based on a bilinear shear stress-slip relationship for concrete and both pre- and post-cracking behaviour was taken into account. The mode II-fracture energy G_f is defined as the area under the τ_l-s_l curve and is used to calculate the maximum transferable load. The anchorage length l_a is the length needed to attain 97 % of this maximum load. In the ultimate limit state the following equations can be found:

Maximum transferable load:

$$P_{max} = b_l \sqrt{2\, G_f E_l h_l} \tag{9}$$

Anchorage length:

$$l_a = \frac{2\lambda + \text{Atan}\left(\dfrac{0.96}{\lambda}\right)}{\lambda\,\omega} \tag{10}$$

with, b_l width of the (CFRP) laminate (mm)
 h_l thickness of the (CFRP) laminate (mm)
 G_f fracture energy (N/mm) ($G_f = (\tau_{lm}\, s_{lm})/2$)
 E_l modulus of elasticity of (CFRP) laminate (N/mm^2)
 E_c modulus of elasticity of concrete (N/mm^2)

$$\lambda = \sqrt{\frac{s_{lm}}{s_{l0} - s_{lm}}}$$

$$\omega = \sqrt{\frac{\tau_{lm}\,(1 + m_l\gamma_l)}{s_{lm}\, E_l\, h_l}}$$

 τ_{lm} maximum shear stress (N/mm^2)
 s_{lm} slip at maximum shear stress τ_{lm} (mm)
 s_{l0} slip where shear stress becomes zero (mm)
 m_l the stiffness ratio ($m_l = E_l / E_c$)
 γ_l the cross sectional area ratio ($\gamma_l = A_l / A_c$)
 A_l cross sectional area of (CFRP) laminate (mm^2)
 A_c cross sectional area of concrete (mm^2)

The model parameters τ_{lm}, s_{lm} and s_{l0} are simply derived from the concrete compressive strength and the concrete tensile strength. Therefore 24 shear tests were performed. The effects of the epoxy resin, the bonding length and the laminate width are examined. The results of these experiments are presented in [11]. The fracture load can be predicted very well which proves the validity of the assumptions and simplification in the model. Since the model is based on simple shear test cases, an extension to combined bending/shear load is needed. Further research on this topic is still going on.

CONCLUSION AND RECOMMENDATION

The technique of externally bonded CFRP laminates achieves considerable strengthening efficiency, particularly when the proposed techniques are provided at the end region of the longitudinal laminates. The authors also recommend providing the first proposed system, but if there is any hindered condition one should execute the second proposed system in order to reach the maximum exploitation of the used fiber.

The experiments allowed to set up a modified design formula for the load carrying capacity of CFRP strengthened r.c. beams, in case of failure due to laminates separation or laminates end shear. The correspondence between experiments and predicted values is superior to the predictions by formulae available in literature up to now.

Further research is needed and will be continued to generalize the proposed equation, and to take into account the different layouts of the CFRP reinforcement, the original strength of the beam to be strengthened in term of the ratios of both tension and shear reinforcements, the shear span to depth ratio, the strength of concrete.

ACKNOWLEDGEMENT

The research is sponsored by the Flemish institute for the promotion of scientific and technological research in the industry (IWT). The Egyptian government also supported this research according to its plan to promote the modern technology in all aspects.

REFERENCES

1. VAN GEMERT, D. Special design aspects of adhesive bonding plates, ACI SP 165 Repair and Strengthening of Concrete Members with Adhesive Bonded Plates, Ed. N. Swamy and R. Gaul, 1996, pp. 25-42.

2. JANSZE, W. Strengthening of R.C. Members in Bending by Externally Bonded Steel Plates Ph. D. thesis, Delft University of Technology, Delft, 1997.

3. BROSENS, K, and VAN GEMERT, D. Plate end shear design for external CFRP laminates, Third International Conference on Fracture Mechanics of Concrete and Concrete Structures, FRAMCOS-3, Gifu, Japan, October 12-16, 1998.

4. VAN GEMERT, D, and AHMED, O. Enhancement of R.C. Beams by Means of Externally Bonded CFRP Laminates, 8th International Colloquium on Structural and Geotechnical Engineering (8th ICSGE), Cairo, 15-17 December, 1998, Egypt.

5. BROSENS, K, and VAN GEMERT D. Stress analysis in the anchorage zones of externally bonded CFRP laminates, International conference on Infrastructure regeneration and Rehabilitation - Improving the quality of life through better construction - A Vision for the next Millennium, Sheffield, England, June 28 - July 2, 1999.

ON THE ANCHORAGE TO CONCRETE OF SIKACARBODUR CFRP STRIPS: FREE ANCHORAGE, ANCHORAGE DEVICES AND TEST RESULTS

A P Jensen   T Thorsen
Technical University of Denmark
E Poulsen   C Ottosen
AEC Laboratory   Sika Beton A/S
Denmark
C G Petersen
Germann Instruments Inc
United States of America

ABSTRACT. The paper describes laboratory tests carried out in order to determine the ultimate force that a CFRP strip is able to obtain when bonded to concrete. Test series have been carried out for varying values of parameters such as concrete strength, anchorage length, thickness of the adhesive, and special anchorage devices like bonded transverse strips and bolted anchor plates. The bonded transverse strips and the bolted anchor plates have been tested in various numbers and positions across the anchorage length. The ultimate free anchorage force has been tested in relation to concrete strength (determined by LOK-test, CAPO-test, BOND-test and TORQ-test) including various types of SikaCaboDur CFRP strips (types S, M, and H). The mechanism of failure of free anchorage is explained and compared with the test observations. The test observations are also compared with the German recommendation for the design of the strengthening of structural components with SikaCarboDur CFRP strips. The structural behaviour of the anchorage devices is explained and compared with observations made during the laboratory tests in order to develop a mathematical model of the anchorage of CFRP strips bonded to concrete. On the basis of the tests carried out, it is discussed how the anchorage devices could be improved in future test series, and if the test set-up is relevant to the problem.

Keywords. CFRP strips, Free anchorage, Anchor devices, Bonding agents, Failure criterion, Test methods, Mathematical models.

Aage P Jensen MSc CEng, PhD, is an Associate Professor at the Technical University of Denmark, IABM, Denmark.

Claus Germann Petersen BSc Mech Eng, MSc econ is the President of Germann Instruments Inc, United States of America.

Ervin Poulsen is Professor Emeritus of CEng AEClaboratory A/S, Denmark.

Christian Ottosen BSc, is the Manager of Technical Services at Sika-Beton, Denmark.

Torsten Thorsen BSc Ceng is an Associate Professor at the Technical University of Denmark, IABM, Denmark.

INTRODUCTION

Since the SikaCarboDur system of strengthening structural concrete components by epoxy-bonded strips of Carbon Fiber-Reinforced Polymers (CFRP) was presented in Denmark several test series have been carried out at university level.

The laboratory tests carried out have dealt with:

Failure conditions of the epoxy adhesive Sikadur-30, cf. [1].

- Simply supported slender beams strengthened by SikaCarboDur CFRP strips.

- Simply supported deep beams strengthened by SikaCarboDur CFRP strips.

- Anchorage tests of CFRP strips epoxy-bonded to concrete specimens applying the system SikaCarboDur, cf. [2].

- Simply supported slender beams strengthened by the system SikaCarboDur, with the application of special made fixing steel plates & bolts and transverse CFRP strips to increase the ultimate anchorage force, cf. [3].

The results of these laboratory tests have clearly shown that anchorage of SikaCarboDur CFRP strips can be improved significantly by the supplement of special inserts (anchor bolts) which act together with the epoxy-bonded CFRP strips in a similar way as the main reinforcement and the stirrups (i.e. links) of an ordinary reinforced concrete beam.

These findings are neither new nor special. Laboratories in the UK and Germany have also found that the ultimate anchorage force of CFRP strips epoxy-bonded to the concrete do not always balance the extreme high tensile strength of the strips. Anchor bolts have always been applied to epoxy-bonded steel plates in the past. However, the application of such bolts to the fixing of CFRP strips ought to be specially designed in order to overcome the anisotropic strength behaviour of the CFRP strips.

FREE ANCHORAGE OF A CFRP STRIP

The Anchorage Problem

From ordinary reinforced concrete it is well-known that a necessary condition for a sufficient anchorage of a reinforcing bar in a concrete member is the presence of a reaction force and / or stirrups in the shape of (rectangular) links, i.e. transverse reinforcement. The purpose of such a structural arrangement is to hold the main steel together, to avoid splitting, and to be able to carry the tensile force into the compression zone to balance equilibrium, cf. Figure 1.

When no reaction force or stirrups are present at the end of the concrete beam to squeeze and hold the reinforcement, the beam will have a tendency to split. The ability of the concrete to resist tensile stresses is rather limited especially when the beam is supported at its end by a corbel, cf. Figure 1.

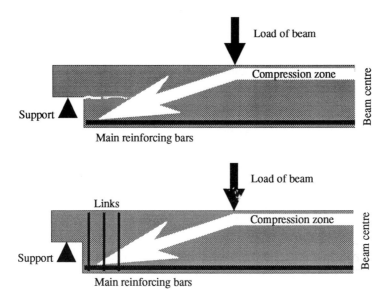

Figure 1   Splitting of a corbel supported beam with insufficient links at the end of the beam
due to the inability of the beam to carry the load induced by arch action

The Anchorage Length of a CFRP Strip

Theoretical considerations have shown that the end of an anchored CFRP strip is subjected to a splitting force or peeling stresses, cf. [4]. Tests carried out in order to determine the ultimate force of an anchorage have shown that splitting or peeling are common modes of failure.

Recent tests carried out at the IABM-Laboratory of the Technical University of Denmark showed that, generally speaking, the ultimate anchorage force of a CFRP strip, epoxy-bonded to concrete, is independent of the length glued to the concrete as long as the glued length of the CFRFP strip is longer than approximately 300 mm (for a medium-strength concrete). The anchorage length is here defined as the length of a CFRP strip beyond a crack of a concrete member, e.g. a flexural or a shear crack.

The German Recommendation for the Anchorage of SikaCarboDur CFRP Strips

Assuming the concrete, epoxy adhesive and the concrete to behave linear elastically O Volkersen established an equation to solve the anchorage problem, cf. [5]. The complexity of this equation has been beaten by a simple approximation developed by the German Institute of Construction Technology, cf. [6]. Here, the ultimate anchorage force F_u (kN) is approximated to:

$$\frac{F_u}{F_{max}} \approx \begin{cases} \dfrac{l_f}{l_{cr}}\left(2-\dfrac{l_f}{l_{cr}}\right) & \text{for } l_f \le l_{cr} \\ 1 & \text{for } l_f \ge l_{cr} \end{cases} \tag{1}$$

where,

$$l_{cr} \approx 0.7\sqrt{\frac{t_f\,E_f}{f_a}} \qquad \text{[mm]} \tag{2}$$

$$F_{max} = 0.5 \times 10^{-3} \times b_f k_b k_T \sqrt{t_f E_f f_a} \quad \text{[kN]} \tag{3}$$

$$k_b = 1.06\sqrt{\frac{2 - b_f/b_{ef}}{1 + b_f/400}} \tag{4}$$

f_a = the mean pull-off strength of the (cleaned) concrete surface [MPa]. The pull-off strength of the substrate concrete should be at least 1.5 MPa and maximum value to apply in the formulae Equation (1) to Equation (3) is 3.0 MPa.

E_f = the elastic modulus of the CFRP strip [MPa].
l_f = the actual anchorage length of the CFRP strip [mm].
t_f = the thickness of the CFRP strip [mm].
b_f = the width of the CFRP strip [mm].
b_{ef} = the effective width of the concrete substrate, e.g. the width of the beam or the distance between the strips glued to a concrete slab [mm].
k_T = a factor depending on the temperature of the structure. from 0.9 to 1.0.

Tests on Free Anchorage of CFRP Strips

To test the ultimate anchorage force of CFRP strips epoxy-bonded to a concrete substrate a specimen of the size 600 x 600 x 100 mm has been used. To each concrete specimen four sets of SikaCarboDur CFRP strips were epoxy-bonded applying the adhesive SikaDur-30. Each set included two identical strips. The in-situ strength of the concrete was tested including three numbers of each of LOK-test, CAPO-test, pull-off test and TORQ-test, cf. [7] and [8].

Concrete Compressive Strength

Twelve concrete specimens of the size 600 x 600 x 100 mm were cast having a compressive strength from 30 MPa to 125 MPa. The in-situ compressive strength of the concrete was determined by means of LOK-tests and CAPO-tests, three observations of each per specimen. This also made it possible to plot the CAPO-strength of the concrete versus the LOK-strength of the concrete. From this it is seen that there is no significant difference between the LOK-test and the CAPO-test. Thus, in the following the compressive in-situ strength of the concrete is defined as the average of the LOK-strength and the CAPO-strength.

Concrete Tensile and Shear Strength

The in-situ tensile strength of the concrete was determined by means of pull-off tests, three observations per specimen. The in-situ shear strength of the concrete was determined by means of the TORQ-test, cf. [7] and [8], three observations per specimen. It is normal practice to relate the tensile strength ft and the shear strength f_v to the compressive strength of the concrete f_c. This made it possible to plot *the f_v/f_c* versus *the f_t/f_c*. These tests illustrate that the in-situ shear strength is approximately 1.8 times the in-situ tensile strength. However, the number of observations is too small to give a more detailed relation. The specimens used in these tests are without any cracks and delamination. For and old concrete substrate with cracks and delamination it is believed that delaminates and cracks perpendicular to the surface have different influences on the tensile strength and the shear strength. Thus it is believed that the in-situ shear strength of the concrete fV, measured by the TORQ-test, is more relevant to the CFRP strip anchorage than the in-situ tensile strength of the concrete ft measured by the pull-out test. However, these tests are not able to prove this hypothesis.

CFRP Strips

The ultimate anchorage force of each SikaCarboDur CFRP strip was determined by means of the DSS-test, cf. [8]. The influence of special parameters such as the thickness of the adhesive joint was included in the tests since this parameter does not appear in the German recommendation for the calculation of the ultimate anchorage force.

Comparison with German Recommendation

A main purpose of the tests was to compare the observed ultimate anchorage force with the calculation according to the strength property of the concrete substrate, the geometry of the SikaCarboDur CFRP strips and its elastic and strength properties. Figure 2 shows the observed ultimate anchorage forces versus their respective calculated values according to the German recommendation of the anchorage of SikaCarboDur CFRP strips.

Figure 2 is supplied with observations made by Bjorn Taljsten, cf. [4]. These plots are interesting since the tests are carried out testing e.g. steel plates and GFRP strips. The strength of the concrete substrate was constantly 25 MPa. As seen, calculation of the ultimate anchorage force could be estimated by the German recommendation.

CFRP STRIP WITH ANCHOR DEVICES

The CRFP Strip-to-Steel Connection

CFRP strips epoxy-bonded to an existing concrete beam or slab will not be able to be squeezed by the reaction force as previously explained. However, by applying a special device (e.g. plate and bolt), cf. Figure 3, the diagonal compression force (arch action) can be balanced by tension in the bolt.

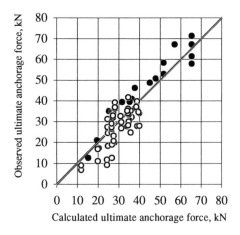

Figure 2 Observed and calculated ultimate anchorage force of SikaCarboDur CFRP strips (O) and various steel plates and GFRP strips (●) according to German recommendations

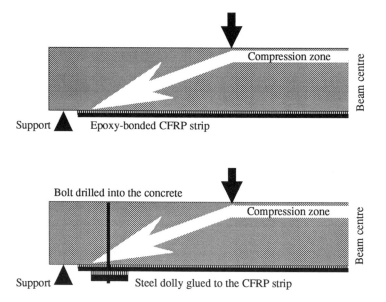

Figure 3 Principle of loading concrete specimen when ultimate anchorage force is determined for a CFRP strip epoxy bonded to the concrete test specimen

Without the plate and bolt shown, anchorage of the CFRP strips must rely on the adhesive agent and the tensile strength of the concrete, cf. Figure 3. When a CFRP strip is epoxy-bonded to a concrete surface and loaded until failure it is observed that the adhesive (Sikadur-30) is so strong that a sliding failure will always occur in the concrete upon reaching the ultimate stage.

In the case of a CFRP strip glued to a steel plate, a sliding failure will occur in the interaction zone between the steel plate and the adhesive. Until now it has not been possible to make the surface of the steel plate rough enough to ensure a sliding failure within the layer of Sikadur-30. However, it has been possible to achieve a sliding failure within the matrix of the CFRP strip.

Improvements of Ultimate Anchorage Force

In order to improve the ultimate anchorage force of a CFRP strip epoxy-bonded to concrete several methods seem possible:

- Glue the CFRP strip at its ends to inserts of steel plate fixed or cast into the concrete.
- Glue steel dollies to the CFRP strips and fix the CFRP strips after their gluing to the concrete by bolts drilled and fixed into the concrete.
- Glue transverse CFRP strips at the end of the longitudinal CFRP strip.
- Glue a CFRP anchor plate at the end of the longitudinal CFRP strip. The optimal shape of such CFRP anchor plates may be inspired by the traditional steel trusses having welded plates at its joints. These anchor plates could be made of steel, but the new SikaCarboDur Wrap might be suitable for this purpose.

Tests have been carried out at the Laboratory of Aalborg University Centre, and at the IABM Laboratory of The Technical University of Denmark. It has been studied if the idea of applying steel dollies glued to the CFRP strips would make it possible to increase the ultimate anchorage force sufficiently.

Laboratory tests have shown that the system of steel dollies and bolts described is a very effective way of increasing the ultimate anchorage force of a CFRP strip epoxy-bonded to a concrete beam, i.e. 'epoxy-bolted'. The tests have been carried out by using steel dollies. These dollies have to be protected against corrosion if used out of doors in practice (stainless steel is not allowed in plate bonding). Thus, one of the main ideas of using epoxy-bonded CFRP strip fails (i.e. to avoid protection against corrosion). Therefore, if such dollies are to be used special CFRP dollies have to be developed, since no holes must be drilled through the SikaCarboDur CFRP strips because it will weaken the anisopropic strip.

The application of 'epoxy-bolted' CFRP strips is time-consuming. Therefore, the idea of using transverse CFRP strips epoxy-bonded across the end of the longitudinal CFRP strips seems better from this point of view, but it has to be remembered that these anchor devices have to rely on the tensile strength of the concrete. According to the Code of Practice in many countries this is not legal or it would require a partial safety factor of the concrete's tensile strength to be considerably larger than the partial safety factor of the compressive strength in order to take care of the brittle failure.

Tests on Anchor Devices

In order to determine the basic properties and characteristics of CFRP strips either anchored by epoxy-bonded 'plates & bolts' or epoxy-bonded transverse CFRP strips it is most convenient to use a test arrangement as shown in Figure 4.

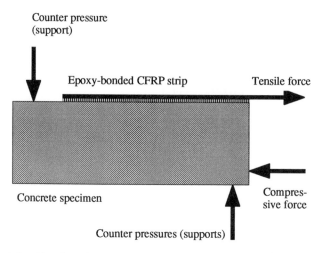

Figure 4 Principle of loading the concrete test specimen when the ultimate anchorage force is
determined for a CFRP strip epoxy bonded to the concrete specimen

'Epoxy-bolted' CFRP Strips with Anchor Plates

A concrete prism with a length of at least 1 m, and a rectangular cross-section of at least 0.1 x
0.2 m in size is suitable as a test specimen. The CFRP strip is 'epoxy-bolted' to the concrete
test specimen and supported as shown in Figure 4. A device for applying a tensile force to the
CFRP strip and counter pressures ensure a statically determinate system of reactions,
cf. Figure 4.

The ultimate anchorage force of the CFRP strip is determined for various numbers and
locations of the anchor bolts. During the loading, measurements of strain are obtained.

There are various ways in which a relation of the ultimate anchorage force of a CFRP strip
'epoxy-bolted' to the concrete test specimen can be obtained. The method of Morsch truss-
analogy has proved its applicability for the design of the stirrups in ordinary reinforced
concrete beams. However, other methods are also possible to apply. The method of Morsch
truss-analogy is simple and is recommended as a first step to a better understanding of the
failure modes of CFRP strips 'epoxy-bolted' to concrete and loaded in tension.

When a formula for the ultimate anchorage force of a CFRP strip 'epoxy-bolted' to concrete is
developed, the design formula ought to be checked by application to simply supported
concrete beams as well as to beams continuously over e.g. two spans. This work has not
started yet. The purpose of testing simply supported beams strengthened by CFRP strips
'epoxy-bolted' to the concrete is mainly to observe if there is a size-effect. The purpose of
testing two-span beams strengthened by CFRP strips 'epoxy-bolted' to the concrete is to
observe if the plastic-hinge theory is valid for the application to statically indeterminate
reinforced concrete structures.

Epoxy bonded CFRP strip with transverse CFRP strips

A concrete prism with a length of at least 1 m, and a rectangular cross-section of at least 0.1 x 1.0 m in size is suitable as a test specimen. The CFRP strip is glued to the concrete specimen and transverse CFRP strips are glued across the main CFRP strip as well. The test specimen is supported as shown in Figure 4. A device for applying a tensile force to the main CFRP strip and counter pressures ensure a statically determinate system of reactions, cf. Figure 4.

The ultimate anchorage force of the CFRP strip is determined for various numbers, Figure 5 and locations of the epoxy-bonded transverse CFRP strips. During the loading, measurements of strain are obtained.

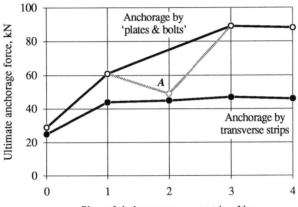

Figure 5 Ultimate anchorage force versus the numbers of
transverse strips or numbers of "plates and bolts" [3]

There is not a simple way in which relations of the ultimate anchorage force of a CFRP strip epoxy-bonded to the concrete test specimen and anchored by means of transverse CFRP strips can be obtained. The method of linear elasticity, cf. [4], combined with the failure criterion of the concrete surface, cf. [7], seems to be the best answer. At least it could be recommended as a first step to a better understanding of the failure modes of a CFRP strip anchored to concrete by transverse CFRP strips epoxy-bonded across the main CFRP strip and loaded in tension.

When a formula for the ultimate anchorage force of a CFRP strip 'epoxy-bolted' to concrete is developed, the design formula ought to be checked by application to concrete slabs. They could be simply supported like balconies, as well as slabs continuously over e.g. two spans. Finally, slabs spanning in two directions, typically used in buildings, must be tested as well. This work has not started yet. The purpose of carrying out tests of simply supported slabs strengthened by CFRP strips epoxy-bonded to the concrete and provided with transverse CFRP strips is mainly to observe if there is a size-effect. The purpose of testing two-way spanning slabs strengthened by CFRP strips epoxy-bonded to the concrete and placed perpendicular to each other is mainly to observe if the yield line theory is valid for slabs strengthened by CFRP strips.

DISCUSSION AND CONCLUSION

This paper describes the need for further investigation of the anchorage problem of epoxy-bonded CFRP strips observed by the first structural tests of SikaCarboDur in Denmark. The paper sketches examples of testing the anchorage of CFRP strips 'epoxy-bolted' to concrete as well as the application of epoxy-bonded transverse CFRP strips without going into details. Finally, the paper suggests that various RC beams and slabs, strengthened by epoxy-bonded CFRP strips are tested under laboratory conditions in order to study the validity of the mechanisms observed. The test set-up used by the tests here reported can be discussed. The testing apparatus is simple in use and the reactions (counter pressures) are statically determinate. However, the strengthened concrete specimens loaded in this testing apparatus, loaded as simply supported beams, and loaded as continuously supported behave structurally in different ways.   Therefore, it is not possible to rely entirely on the results from the observations made using the testing setup described. However, the testing apparatus is simple and it is possible to study the mechanisms of failures.

REFERENCES

1. ERVIN POULSEN, AAGE P. JENSEN, CHR. OrrosEN. The failure criteria for Sikadur-30 and -31 adhesives tested by CEN test method prEN 12188. US-Canada-Europe Workshop on Bridge Engineering, Dubendorf and Zurich, Switzerland 1997.

2. BYTOFT OLSEN, L., GERTSEN L. Styrkemal for betons overfladelag— vedhaftning af kulfiberband (Strength of the near-to-surface layer of concrete - Bond of CFRP strips, in Danish). The Technical University of Denmark, IABM. Lyngby, January 1998.

3. CHRISTENSEN, B., BACHE, H. Forankring af CFRP-lameller (Anchorage of CFRP strips, in Danish). The Technical University of Denmark, IABM. Lyngby, January 1997.

4. BJORN TALJSTEN. Plate bonding, strengthening of existing concrtete structures with epoxy bonded plates of steel or fibre reinforced plastics. Doctural Thesis. 1994:152 D. Lulea University of Technology, Division of Structural Engineering. Lulea, Sweden 1994.

5. VOLKERSEN, O. Die Nietkraftverteilung in zugbe-anspruchten Nietverbindungen mit konstanten Laschenquerschnitten (in German). Luftfartforschung, Vol. 15, 1938.

6. GERMAN INSTITUTE OF CONSTRUCTION TECHNOLOGY. Strengthning of reinforced concrete and prestressed concrete with Sika CarboDur bonded carbon fibre plates. Authorisation No.: Z-36.12-29. Berlin November 1997.

7. CLAUS GERMANN PETERSEN, EKVIN POULSEN. In-situ testing of near-to-surface layer of concrete and epoxy-bonded CFRP strips. Proceedings of the US-Canada-Europe Workshop on Bridge Engineering, Dubendorf and Zurich, Switzerland 1997.

8. CLAUS GERMANN PETERSEN. In-situ test systems for durability, inspection and repair of reinforced concrete structures. Germann Instruments A/S. Copenhagen, Denmark 1998.

STRENGTHENING OF REINFORCED CONCRETE STRUCTURES BY PRESTRESSED OR NON-PRESTRESSED EXTERNALLY BONDED CARBON FIBRE REINFORCED POLYMER (CFRP) STRIPS

H-P Andrä

D Sandner

M Maier

Leonhardt, Andrä und Partner

Germany

ABSTRACT. There are many reasons for post-strengthening of reinforced concrete structures, such as serviceability improvements, increased traffic volumes on bridges, steel reinforcement corrosion, decrease in deformation etc. Strengthening structures by means of bonding steel plates with adhesives has become standard engineering practice since 1967. Steel can be replaced by the light-weight and non-corrosive CFRP-material. General construction authorisation was given to the product Sika CarboDur CFRP strips by the German Institute of Construction Technology. The strips are so far used in non-prestressed condition. In order to improve the serviceability of strengthened structures a site-suitable system has been developed by Leonhardt, Andrä and Partners which allows to apply the CFRP strips in prestressed condition. The worldwide first practical application of post-tensioned externally bonded CFRP strips (LEOBA-CarboDur surface-tendon) was carried out successfully in October 1998.

Keywords: Carbon, CFRP-laminates, Strengthening of structures, Non-prestressed / prestressed externally bonded reinforcement

Dr-Ing MSc H-P Andrä is Managing Director of Leonhardt, Andrä und Partner, Consulting Engineers VBI, GmbH Stuttgart, Berlin, Dresden, Erfurt Germany

Dipl-Ing D Sandner, Research and Development, Leonhardt, Andrä und Partner

Dipl-Ing M Maier, Project Engineer, Leonhardt, Andrä und Partner

INTRODUCTION

Fibre reinforced plastics, a combination of fibres and a matrix, are used as glass-fibre (GFR), aramid-fibre (AFR) and carbon-fibre (CFR) reinforced materials.

Externally bonded Carbon-fibre reinforced polymers (CFRP) laminates are particularly suitable for post-strengthening and structural repair of reinforced concrete structures.

In 1996, Leonhardt Andrä and Partners in collaboration with the building supervisory authority could use this strengthening method in Baden-Württemberg for the first time. A new storey was built on top of the existing roof - a two-way ribbed floor structure - of a building at Stuttgart University. At that time each projected CFRP application had to be approved individually. Since 1997 strengthening of reinforced and prestressed concrete by externally bonded Sika CarboDur CFRP strips is authorised by the German Institute of Construction Technology. Design regulations and guidelines for the strengthening with Sika CarboDur have been defined. The same year the damaged prestressed girders of two church roof structures were strengthened.

While the experience in structural design of "regular" (non-prestressed) CFRP applications grew, it turned out, that the only way to use the full load-bearing capacity of this high tensile material is to apply it in prestressed condition.

Therefore a new prestressing- and anchorage system has been developed by Leonhardt, Andrä and Partner, which provides a maximum prestress of about $1000 \cdot N/mm^2$ (at a strain of 0.6 %) to each laminate, and allows direct anchorage on the surface of the member to be strengthened.

The system can easily be handled on-site and was tested successfully at Swiss Federal Laboratories for Material Testing and Research (EMPA) in Zürich.

The first practical on-site application of externally bonded prestressed CFRP strips was carried out successfully for rehabilitation reasons of a bridge structure in southern Germany in October 1998.

The systems features, the results of the EMPA-test, the final on-site application and the faced problems will be presented and illustrated.

MATERIAL SPECIFICATIONS

Carbon Fibres

The carbon fibres are Toray T 700 SC endless C fibres from Toray Industry, Japan. The fibres' diameter vary from 5 up to 8 μm. (The diameter of human hair ≈ 100 μm).

The mechanical properties are

- Axial tensile strength $\quad\quad\quad\quad\quad\quad\quad f_r \geq \quad 4800 \quad N/mm^2$
- Elastic modulus in direction of fibre $\quad\quad E_f \geq \quad 200,000 \quad N/mm^2$
- Ultimate elongation in direction of fibre $\quad \varepsilon_{fu} \geq \quad 2.0 \quad \%$

Matrix Resin

The resin system is an epoxy resin.

- Tensile strength f_m $\geq$ 70 N/mm^2
- Elastic modulus E_m $\geq$ 3000 N/mm^2
- Ultimate elongation ε_m $\geq$ 4.0   %
- Poisson's ratio v_m $\geq$ 0.3   %

Sika Carbodur Strips

The CFRP strips are pultruded laminates 1.2 - 1.4 mm thick, 70 vol.% reinforced by unidirectional carbon fibres Toray T 700 SC. The strips width may be 50, 60, 80 100, 120 or 150 mm.

- Tensile strength f_l $\geq$ 2800 N/mm^2
- Elastic modulus in direction of fibre E_l $\geq$ 170,000N/mm^2
- Ultimate elongation in direction of fibre (average) ε_{lu} $\geq$ 1,6   %

Epoxy Adhesive

For bonding the CFRP plates with concrete / plates to each other Sikadur 30 is used.

- Time of application   40-80 Min   at 20 ° C
- Stability   3-5 mm   at 35 ° C
- wet adhesion (break in concrete)   4 N/mm^2
- Elastic modulus   8000 – 16000 N/mm^2

ADVANTAGES OF CFRP STRIPS FOR ADHESIVE REINFORCEMENT

- any lengths available
- corrosion resistant
- very high tensile strength
- extremely low weight
- low installation costs, easy joints
- flexible (The minimum radius to be met when transporting the plates in reeled form is 0.90 m)

NON-PRESTRESSED EXTERNALLY BONDED CFRP LAMINATES

Requirements

The use of Sika CarboDur CFRP strips requires at first a solid study of the present state of the member to be strengthened. Concrete strength, surface tensile strength, steel type, position and state of the existing reinforcement as well as position, line and width of cracks shall be recorded. Further requirements for the member to be strengthened, for the steel parts and for the CFRP material are to be defined.

Structural Analyses

At first it should be taken into account, that the stress–strain line of the existing reinforcement can be assumed to be   bilinear with $E_s = 210000$ N/mm^2 ,while that of the CFRP plate is linear.

As the CFRP possesses no yielding deformation reserve, the maximum bending strength of a strengthened section is obtained, if rupture of the CFRP plate occurs once the internal steel reinforcement reached its yield-range and prior to concrete fracture. For this reason and to avoid delamination- and debonding effects the ultimate strain in design of CFRP strips is limited.

1.   *Ultimate strain of CFRP strip for bending design*

      ε_{lu} $\leq 5\ \varepsilon_{sy}$ for reinforcement steel

      ε_{lu} $\leq \varepsilon_{cu}\ /\ 2$ for Sika CarboDur CFRP strip

      $\Rightarrow$ The smaller value is governing

      ε_{lu} = ultimate plate strain for bending design

      ε_{sy} = yield strain of the internal steel reinforcement

      ε_{cu} = ultimate tensile strain of the CFRP plate

A strain diagram is shown on Figure 1. The ultimate strain ε_{lu} in the CFRP strip after strengthening is given here as $\Delta\varepsilon_l$. ε_{s0} defines the steel strain due to existing permanent loads before strengthening. $\Delta\varepsilon_s$ defines the steel strain due to additional loads after strengthening.

The total steel strain is given as

$$\varepsilon_{s0} + \Delta\varepsilon_s\ =\ \varepsilon_s$$

Non-prestressed CFRP laminate

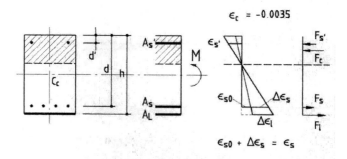

Strain diagram in the ultimate limit state

Figure 1   Strain diagram - non-prestressed CFRP

2.   *Degree of strengthening*

The design bearing capacity of the strengthened member shall not climb over double the amount of that of the unstrengthened one. The degree of flexural strengthening is limited to $\eta_B \leq 2$.

$\eta_B = M_{uV} / M_{u0} \leq 2$

M_{u0} ultimate bending moment of an unstrengthened member

M_{uV} ultimate bending moment of a strengthened member

3.   *Design of adhesive bond anchorage*

The anchorage of a CFRP plate mainly depends on the characteristic bond failure force $T_{k,\ max}$ and the anchorage length $l_{t,\ max}$. In both of these values the characteristic surface tensile strength, the plate geometry , the E-modulus of the CFRP material and the width of the member are to be taken into account. The required anchorage length l_t and the required bond failure force T_k are to be adjusted to the tensile force cover line for the design failure state.

Application

Figure 2 shows the strengthening of an existing roof two-way ribbed floor structure of a building at Stuttgart University in 1996. The bond jointing work at the crossings of this two-way ribbed floor is very easy compared with the application of steel plates.

Figure 2 CFRP strengthened two-way ribbed floor

PRESTRESSED EXTERNALLY BONDED CFRP LAMINATES

Advantages of Prestressed CFRP Laminates

Prestressed CFRP strips are able to participate in carrying dead loads and permanent loads. Prestressing improves the serviceability of the strengthened member. Especially the strains of the internal steel reinforcement will be relieved. The crack width and the bending deflection can be reduced. Further on the load bearing capacity can be increased, as crack widths are reduced and the mutual vertical displacements of crack boundaries impeded. The danger of delamination of the strip due to shear forces can be reduced.

Crack's faces can be taped and tightened together by creating a local compression area. The section can be forced back to the uncracked stage and thus gaining flexural rigidity.

System's Features

The stressing operation is carried out by a hydraulic tandem cylinder miniature press (Figure 3). New end anchors have been developed to solve the peeling and debonding tendencies of a prestressed laminate at its end while exceeding the concrete's tensile strength. The single stages of the prestressing operation and the anchorage system are shown on Figure 4 and 5.

Figure 3 Miniature-press,
performance of 70 kN at 1450 bar   Figure 4   Stressing operation

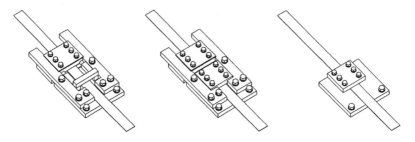

(1) Stressing operation − (2) curing of the adhesive − (3) final permanent stage

Figure 5   Stages 1 − 3 of the anchorage system

Final Test at EMPA Laboratories, Zurich

The prestressing system has been tested successfully in a large-scale-test at the EMPA, the Swiss Federal Laboratories for Materials Testing and Research. The testing arrangement, and the strain gauges' recordings during the stressing operation are shown in Figure·6 and 7.

Figure 6   Final test of the LEOBA CarboDur
post-tensioning system

Figure 7 strain gauges' recordings during
stressing operation (EMPA test report)

As prestressed externally bonded CFRP tendon are meant to get practically applied on
cracked members, the test beam was loaded before until cracks showed up clearly. Additional
strain gauges were installed to watch the laminate's behavior close to those cracks while
bearing cyclic loadings near the ultimate load range. The recordings showed some kind of
tension stiffening effect. The strains right next to the monitored cracks were running behind
those in-between two cracks

First Practical Application of Prestressed CFRP Strips

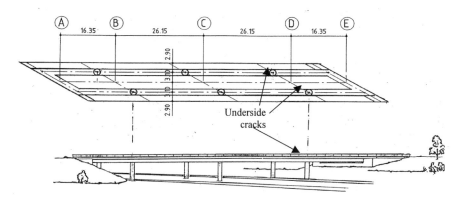

Figure 8 Underside- and sideview of the Lauterbridge in Gomadingen showing wide cracks
right next to the columns' support in the supposed compression area

Damage

The bridge, designed as a 4-span continuous beam with internal bonded tendons individually
for each lane, showed wide cracks at the bottom side of the section, right next to the first
columns' supports.

These cracks, to be found in the supposed compression area of the section, are an obvious indication of a definite tension area and can be explained as follows:

1. The internal tendons were prestressed to a too high degree.

2. The tendons prognosticated decrease of stress, caused by creeping and shrinking, did not occur.

3. The dead load of the concrete structure itself is assumed at a too high rate.

4. The structural affect of the two beams with another – due to the oblique-angled columns'-support – was not considered in design.

Rehabilitation

By applying prestressed externally bonded CFRP strips combined with a resin injection before the cracked section could be forced back to the uncracked stage.

This gain of stiffness improved the serviceability and relieved the strains of the internal reinforcement. The cracks faces were „taped" and fixed together. They won`t „breathe" anymore while bearing cyclic and reversed loading. A non-prestressed application could not have performed that.

Figure 9   Sideview of the Lauterbridge in Gomadingen

Figure 10 4x LEOBA-CarboDUR CFRP tendons at cracked sections

REFERENCES

1. MARTIN DEURING, EMPA, Swiss Federal Laboratories for Materials Testing Report 224, 1993

2. ROSTASY, HOLZENKÄMPFER, HANKERS, Geklebte Bewehrung für die Verstärkung von Betonbauteilen, Betonkalender 1996, Berlin Ernst& Sohn

3. NEUBAUER, Concrete Precasting Plant and Technology BFT 6 / 1998

4. SIKA CARBODUR bonded carbon fibre plate, German Institute of construction technology, Berlin, authorisation No. Z-36.12-29

5. EMPA, Swiss Federal Laboratories for Materials Testing Report·No172745´/2·1998

APPLYING GALVANISED STEEL FIBRES FOR CORROSION-PROTECTION OF REINFORCED CONCRETE STRUCTURES UNDER DE-ICING EXPOSURE

A Keyvani

Tarbiat Moallem University of Tabriz

Iran

N Saeki

Hokkaido University

Japan

ABSTRACT. The main problem of reinforced concrete structures of highways and roads in snowy and glacial region is corrosion of reinforcement due to applying de-icing. By considering many advantages of steel fibres for concrete matrix, interaction between steel fibres and steel bars to inhibiting corrosion of steel bars was studied under an artificial and aggressive de-icing environment. Severe corrosion of bars was observed in non-fibre reinforced concrete even without the occurrence of any crack phenomena in the specimens by a acceleration corrosion test. However, when galvanized steel fibres and steel bars were placed in contact with each other, the galvanic behavior of fibres cathodically protected the steel bars. The non-corrosion of embedded steel bars in a galvanized steel fibre-reinforced concrete matrix indicated that galvanized steel fibres are a good method for inhibiting corrosion of reinforced concrete members.

Keywords: Galvanized steel fibre, Corrosion-protection, Chloride ions, Acceleration corrosion test, Sacrificial anode, Saline solution.

Dr Eng Abdullah Keyvani is assistant professor in the Department of Civil Engineering at Tarbiat Moallem University of Tabriz, Tabriz, Iran. His main research interests include steel fibre reinforced concrete and durability of reinforced concrete. He is the author of several books and papers on steel fibre reinforced concrete and corrosion protection of reinforced concrete.

Professor Noboru Saeki is director of the Advanced Concrete Engineering Laboratory, Hokkaido University, Sapporo, Japan. He is also the chairman of the Advanced Materials Research Committee of the Hokkaido Chapter of the Japan Concrete Institute. His current research interests are in the area of durability of reinforced concrete and the rehabilitation of reinforced concrete by using FRP. He is the author of several books and numerous papers on various aspects of concrete technology.

INTRODUCTION

The natural alkalinity of concrete prevents steel from corroding unless a large amount of chloride ions penetrates to the surface of the reinforcement. If the concrete is not cracked, the concrete cover over reinforcement can provide protection for many years [1]. The effectiveness of concrete protection depends not only on the lack of cracks but also on the porosity of the diffusion rate of the concrete. The principal disadvantages of concrete are low tensile strength and brittleness.

Steel fibre can improve these tensile and brittle properties, especially if the applied fibres can act as a corrosion inhibitor role for reinforcing steel bars in the concrete matrix. This has been confirmed by the results of experiments on the inhibition of corrosion in steel bars by fibres in a ferroxyl transparent gel simulation environment [2]. The corrosion behavior of steel fibres in steel fibre reinforced concrete (SFRC) has not been studied as methodically as that of reinforcing steel or prestressing steel, particularly with regard to the protection of corrosion in reinforced concrete. In general, the findings regarding corrosion behavior of steel fibres have been good. The corrosion of SFRC has been shown to stop at the surface of members, and corrosion of steel fibres inside the concrete matrix has typically not been observed [3].

One of the major advantages of SFRC with regards to corrosion is that the fibres, being non-continuous and discrete, provide no mechanism for the propagation of corrosion activity. This phenomenon has been clarified from the examination of numerous SFRC and steel fibre reinforced shotcrete structures subjected to aggressive exposure environments [4, 5].

A new method that uses galvanized steel fibre-reinforced concrete has been proposed as a means to inhibit the corrosion of reinforcing steel bars. In this method, galvanized steel fibre (GaSF) galvanically control the corrosion of reinforcing steel bars. Experiments conducted on fibrous concrete under severe environmental conditions in the laboratory have shown that galvanized steel fibres in contact with steel bars act as a sacrificial anode, cathodically protecting embedded steel bars. Laboratory experiments have also shown that GaSF have good resistance to corrosion and depth of corrosion in the fibres limited to 5-8 mm from directly exposed surfaces after 9 months of exposure. Corrosion and dissolution of corrosion products occurred only in fibres that were on exposed surfaces, and no spalling or crack phenomena were observed on the interface of concrete and fibres. In this study, the behavior of galvanized steel fibres under conditions of accelerated chloride exposure and the inhibitory effects of GaSF on corrosion of reinforcing bars were investigated.

EXPERIMENTAL PROGRAMME

The capability of galvanized steel fibres to protect steel bars against corrosion was tested in two groups of experimental reinforced concrete specimens.

Mix Proportions & Specimen Preparation

The first group was reinforced concrete beam specimens of 10x10x40 cm without any fibres. The second group was as same as the first group but with galvanized steel fibres (GaSF). The first group served as control specimens in order to compare the effects of galvanized steel fibres under an aggressive environment. The amount of GaSF was 1.5% by volume of fibres or 120 kg/m^3 and the specifications of applied fibres is shown in Table 1.

Table 1 Specification of galvanised steel fibre (GaSF)

	CHEMICAL PROPERTIES					YIELD POINT STRENGTH (N/mm²)	UNLTIMATE TENSILE STRENGTH (N/mm²)	ELON-GATION (%)	Zn COATING (gm/m²)
SIZE (mm)	C	Si	Mn	P	S				
0.5x0.5x30	<0.12	-	<0.5	<0.04	<0.05	166	>270	>32	120

To accelerate electrolytic corrosion, 3 kg/m³ NaCl in group I and 0.3 kg/m³ NaCl by weight of concrete in group II were added to the concrete mixture during mixing. The mix proportions used are summarized in Table 2. In the beam specimens, two profiled-prestain steel bars (10 mm, 35 cm in length) were embedded at a cover thickness of 16~20 mm. These were connected to lead wires to measure the electric potential of bars at desired times.

For GaSF specimens, all fibres were space-oriented (randomly) around the steel bars and within the matrix. Preparation of mixtures, casting, consolidation of specimens, measuring of air content, slump, and curing were performed in accordance with JSCE-F503-1990, F551-1983, F552-1982, and JSCE Specifications Codes [6]. All specimens were left inside the mold for 24 hours after casting and then stripped off the mold and water cured at 21±2 . for 27 days before being put in an automatic wet-dry artificial aggressive environment apparatus.

Table 2 Mix proportions of the two groups of concrete specimens

GROUP	SAND (kg/m³)	GRAVEL (kg/m³)	CEMENT (kg/m³)	WATER (kg/m³)	W/C	AEA* (l/m³)	STEEL FIBRE (kg/m³)	INITIAL SALT (kg/m³)
SF-14-3	832	1045	280	154	0.55	0.22	120	3.0
SF-17-3	832	1045	280	154	0.55	0.22	120	0.3

* AEA = Air entraining agent

Accelerated Exposure Environment

An artificial wet-dry corrosive environment was considered in accordance with ASTM B-117 [7], with necessary modifications to the reinforced concrete specimens. The wet-dry cycles consisted of 12 hours wet and 12 hours dry every operational day.

The wet portion used a flowing aerated recirculation of 5% sodium chloride aqueous solution, which was showered over the surface of each specimen. The solution had an average temperature of 35. and initial pH of 6.8. In the dry cycle, 35. air was circulated over the specimens.

Variations in the potential of embedded steel bars in both groups were monitored periodically with respect to the silver-silver chloride electrode and then converted to the copper-copper sulfate electrode (CSE) according to ASTM C 876 [8].

RESULTS AND DISCUSSION

Analysis of Chloride Ions Diffusions

Crushed pieces of 28-days-old compressive specimens that had been cured in fresh water were selected for determination of the initial chloride amount. Pieces of the concrete cylinders were divided into two size groups, 0.0~20 mm and 20.1~50 mm (Fig. 1). However, in the beam specimens, chloride analysis was carried out after 90, 180 and 270 days of exposure. To drill out at least 10~15 gr. of powder from the specimens at preselected 5 mm intervals from the exposed surfaces, a carbide-tipped drill of 10 mm in diameter was used.

The level of chloride ions necessary to initiate corrosion has not yet been accurately determined. Some investigators have suggested that the critical level is also dependent on the hydroxide level in the pore solution. The Federal Highway Administration (FHWA) of America has set chloride levels at which a bridge deck must be replaced. These standards are such that when the chloride level is 0.15% by weight based on the cement content, there is no danger of corrosion at that time and the deck can be left intact [9]. When the chloride level reaches 0.3%, the concrete must be removed to below the rebar level or the deck must be replaced completely.

A state-of-the-art survey on this problem prepared by a group in Europe concludes that a consensus on the permissible level of chloride ions in concrete mix is difficult [10]. This same report stated that corrosion will probably occur when the chloride ion concentration reaches 0.35 to 1% by weight based on the cement content.

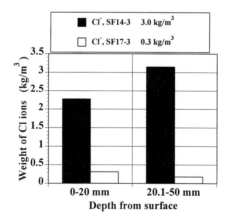

Figure 1 Initial amount of Cl ions after 27-day curing

Results of initial chloride ions in 1. m^3 of concrete are shown in Figure 1. In both the groups of non-fibre and GaSF reinforced specimens, the steel bars were in concrete with chloride concentrations well above the threshold level of 0.3% by weight based on the cement content. The results showed that the rate of Cl^- diffusion at the level of the steel bars for in both either SF17-3 and PC17-3 group II specimens during 3 months and 6 months of exposure were higher than that of group I specimens.

It seems that, for concrete with a low initial addition of Cl^-, the speed of diffusion of Cl^- is faster than that in concrete with a high initial addition of Cl^-. On the other hand, de-icing of reinforced concrete structures in the first winter of servicing would cause a high level of Cl^- diffusion. At 9 months of exposure, the rate of Cl^- diffusion at the level of steel bars was almost same for both groups, Figure 2.

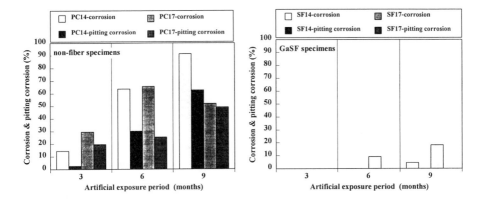

Figure 2   Diffused Cl ions after 9 months exposure for (a) 3 kg/m^3 and (b) 0.3 kg/m^3 initial addition of salt in the specimens of group I and II

Analysis of Cl^- ions in specimens even after 3 months of exposure showed that amounts of existing and diffused Cl^- ions at the surface of GaSF specimens were much higher than those in non-fibre specimens. This phenomenon was thought to be due to the presence of sufficient oxygen and moisture, which caused corrosion activity of galvanized fibres at a depth of 0.0~10 mm. Therefore formation of soluble corrosion products with components of chloride ions and diffusion them into concrete matrix increased Cl^- ions at those layers. Consequently, the rate of diffusion into layers below 10 mm decreased sharply.

Potential Analysis of Embedded Steel Bars

The potential of steel reinforcement can be used to assess the probability of corrosion at the time of measurement. It should be noted that such half-cell potential data can only be used to assess the probability of corrosion and not the rate of corrosion. This means that the results are only qualitative, without establishment of an actual rebar corrosion rate. At times, very negative rebar potential readings can be misleading and may be recorded in submerged concrete, where the corrosion rate remains negligibility low due to unavailability of oxygen stifling the cathodic reaction rate [11].

The electrode potentials of embedded steel bars in non-fibre and GaSF concrete specimens were initially monitored at 28 days and then at 2, 3, 5, 6, 7, 8 and 9 months of exposure and at 2-cm intervals. In all specimens, the initial potential at 28 days after curing in the fresh water became more negative. During the 27-day curing in water, the half-cell potential was more negative due to a lack of oxygen in the specimens.

After one month of exposure, the potentials had moved in the positive direction, which confirmed passivation of the bars. However, the passing of the time, effects of the initial addition of salt, conditions of the wet-dry cycles of the corrosion acceleration apparatus, and the penetration of additional chlorides up to surface of the steel bars facilitated electrochemical reactions and a current between the steel bars and measured surfaces, which resulted in negative potentials. The potentials of non-fibre/PC14-3 specimens in group I (Fig. 3-a) compared to GaSF/SF14-3 specimens showed a big difference between the first and second months of exposure (Fig. 3-b). However, the variation in the of potential of galvanized steel fibre specimens was less than that of non-fibres specimens. On the other hand, in PC14/3 specimens of group I, potentials moved in the negative direction, with values of -400 to -500 mV (CSE). These potentials were more than the threshold level (-350 mV), and therefore there was more than 90% probability of corrosion occurring, which was confirmed by visual observation after breaking the specimens.

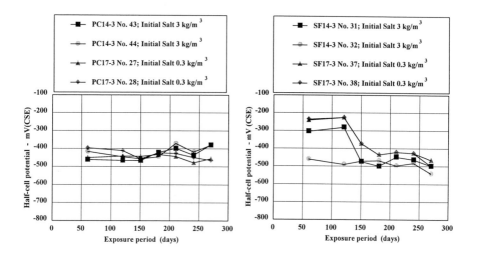

Figure 3 Half-cell potential variation during 2~9 months exposure in the specimens of group I and II a) non-fibre-reinforced specimens, b) GaSF-reinforced specimens

The maximum potential of GaSF/SF14-3 specimens in group I at the end of 3 months and 6 months of exposure were about -550 mV (CSE). It seems that, the potential of contacted galvanized steel fibres with steel bars and the corrosion of them were effective in the measured potentials (Fig. 3-b). Therefore, potentials of more than -350 mV (CSE) with the meaning of more than 90% probability of corrosion occurring concluded. However, the results of corrosion occurring in GaSF specimens could not be correct because after GaSF specimens were broken, non-corrosion of steel bars observed. It means that, a higher potential result than ASTM recommendations was due to the effect of zinc corrosion of galvanized fibres which was in contact in steel bars embedded in a galvanized steel fibre concrete matrix.

Revocation of potentials in the non-fibre specimens after 6 months and 9 months of exposure seems that the chlorides that had penetrated between the steel bars and applied surface might be played as an isolated layer. However, the bars in the galvanized steel fibres matrix showed an uniform variation after each month of exposure.

This uniform variation in the potential of embedded steel bars in contact with galvanized fibre is the main reason for the anodic activity of connected fibres and the chemical reactions of fibres with diffused Cl^-. It can also be assumed that the connected fibres were as a body of steel bars and that the dissolution of the zinc layer caused an increase in the measured potentials. However, the distribution of potentials along the bars in the GaSF specimens were more stable than those in non-fibre concrete specimens.

In the non-fibre/PC17-3 specimens of group II, there was little difference in the variations of potential of 2 or 3 months of exposure, and the potential was more negative than that of non-fibre/PC14-3 specimens of group I (Fig. 3-a). It seems that in non-fibre concrete, the effect of 0.3 kg/m^3 of initial salt for corrosion acceleration was more than 3.0 kg/m^3 .

The results of GaSF/SF17-3 specimens of group II after 3 and 6 months of exposure were more positive than those of GaSF/SF14-3 specimens of group I (Fig. 3-b). However, the potentials after 9 months of exposure in the GaSF specimens of both groups were similar. This was thought to be due to the rate of the initial addition of salt. That is, the high level of the initial addition of salt in the specimens of group I (about 10- times that of group II) might attack the coating of GaSF and steel bars before and during the formation of a protective layer. Therefore, the cathodic protection role of galvanized fibres for the steel bars in group I was less than that in group II. This is because the chlorides that penetrated later from the outside due to exposure had encountered protective layers on fibres and steel bars and therefore may be less aggressive. It is concluded that an initial addition of 0.3 kg/m^3 of salt for corrosion acceleration in non-fibre specimens has a negative potential than 3.0 kg/m^3 initial addition of salt.

However, the behavior was different in GaSF specimens. The main reason for this different behavior is the corrosion activities of GaSFs near or over the steel bars. That is, corrosion of fibres caused a decrease in the amount of oxygen, and the steel bars were thus put in a stable state of passivation.

A shift in half-cell potentials from negative to positive is indicative of a low probability of corrosion, most likely as a result of the formation of a stable passive oxide film on the surface of the specimens in the highly alkaline environment of concrete. The initial rapid increase in the potential is consistent with the idea that initial oxide formation, i.e., when the oxide thickness attains a few atomic layers, is the most important stage of the process [12]. Further thickening of the oxide layer occurs over several days. However, this only results in minor potential changes. Over a long period, an equilibrium is reached between oxide formation and oxide dissolution, thereby resulting in a stable potential.

The potential distribution on the electrode surface of the galvanic corrosion of zinc under a thin layer of electrolytes was measured experimentally for a galvanic couple [13]. The results showed that the galvanic current increases with increases in the area of steel up to a certain size and then decreases slightly with larger areas. It decreases sharply as the distance between zinc and steel increases because the system becomes ohmically resistance-controlled. It is less sensitive to the variation of electrolyte layer thickness.

The width of zinc has little effect on the galvanic current because most anodic reactions take place *in a very narrow area at the edge closest to the steel*. In fact, this behavior is the same as that of galvanized steel fibres on steel bars that they are in contact with in a concrete electrolyte environment

Visual Observation

After 3, 6 and 9 months of exposure by the initial addition of chloride salt and the effects of a severe artificial aggressive environment, the concrete specimens were broken open, and the embedded steel bars were examined visually for extent of coverage by corrosion products, nature of products, behavior of galvanized steel fibres, etc. The following points were observed in non-fibre reinforced concrete specimens and galvanized steel fibre-steel reinforced specimens.

Behavior of steel bars in non-fibre-reinforced concrete beams

Crack phenomena on exposed surfaces and other surfaces of non-fibre specimens, even after 9 months of exposure under aggressive conditions, were not observed. In the all specimens of non-fibre reinforced concrete beams after 3, 6 and 9 months of exposing, serious corrosion in the form of black-colored and/or brown-colored corrosion products were observed with 0.1~0.2 mm thickness. In the all specimens, besides of corrosion, pitting corrosion were occurred and depths of pitting were about 0.4~0.6 mm (Figure 4-a). After inspection of broken specimens, the real depth of cover over the steel bars was measured, which was 16~20 mm cover.

Behavior of steel bars in galvanized steel fibre-reinforced concrete beams

The exposed surfaces and other surfaces of all GaSF specimens were corroded. Orange-colored products, due to rust and corrosion of fibres, covered the surface of beams. The galvanized steel fibres connected to steel bars were partially corroded, and the rate of corrosion was higher in group I specimens, in which the initial amount of added salt was higher. Brown-colored and white-colored corrosion products were clearly observable on the fibres. Corrosion phenomena were observed more in groups of fibres than in scattered fibres. This phenomenon is an important reason for the electrochemical reaction of GaSF within the chloride concentrated matrix or around the steel bars and shows the sacrificial role that GaSF plays to prevent corrosion of steel bars (Figure 4-b).

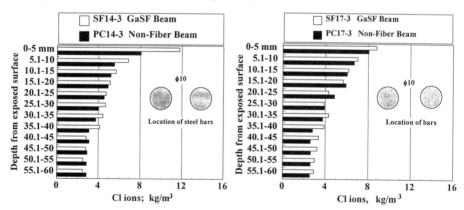

Figure 4 Corrosion and pitting areas in the specimens of group I and II after 3, 6 and 9 months exposure a) non-fibre-reinforced specimens, b) GaSF-reinforced specimens

Despite the severe corrosion of fibres and the spread of corrosion products on exposed surfaces of the beams, close inspection after scratching off the corrosion products showed that there were no cracks around the fibres and surfaces of concrete even after 9 months of exposure. The thickness of corrosion products of GaSF on the exposed surfaces was 0.05~0.1 mm; however, for fibres inside the beams, these products were completely superficial.

After 9 months of exposure, non-fibre specimens showed severe corrosion and pitting phenomena on 16~30% of the surface of bars. In the GaSF specimens of both groups, corrosion occurred superficially as small spots of corrosion products. Consequently, compared to that in non-fibre specimens, the corrosion of steel bars in GaSF specimens was negligible.

CONCLUSIONS

Corrosion behavior of steel bars and protection of steel bars by galvanized steel fibres in a concrete matrix under an acceleration test environment were investigated. The following results were obtained.

1. The volume expansion force of a galvanized steel fibre due to corrosion products is negligible compared to common steel bars, and therefore no spalling phenomena of concrete were observed around the corroded fibres.

2. Fibres in galvanized steel fibrous concrete, either in contact with or placed near steel bars, considerably decreased the formation and propagation of corrosion zones in the steel bars. Corrosion phenomena was not observed in most of the steel bars after 9 months of exposure under severe conditions. However, under the same conditions, steel bars in non-fibre concrete were seriously corroded.

3. Corrosion phenomena in steel fibres which are non-continuous and discrete in the concrete matrix provides no mechanism for the propagation of corrosion activity and spalling of concrete or de-bonding of concrete from steel bars.

REFERENCES

1. KOSMATKA, S H. Concrete Technology Today, No. 8, 2 , June 1988, p. 1.

2. KEYVANI SOMEH, A, SAEKI N, AND NOTOYA, T, Simulation of Role of Steel Fibres For Corrosion Protection of Reinforced Concrete Members, Japan Cement Association Proceedings of Cement & Concrete, No. 50, 1996, p.53.

3. KEYVANI SOMEH, A, SAEKI, N AND NOTOYA, T, Corrosion of Steel Fibre Reinforced Concrete, Fourth ACI/CANMET International Conference on Durability of Concrete, Australia, 1997, Supplementary Papers, pp. 451-466.

4. STARK, D, Measurement Techniques and Evaluation of Galvanized Reinforced Steel in Concrete Structures in Bermuda, Corrosion of Reinforcing Steel in Concrete ASTM STP 713, D. E. Tonini and J. M. Gaidis, Eds., ASTM, 1980, p. 132.

5. MORGAN, R, Steel Fibre Reinforced Shotcrete for Support of Underground Openings in Canada Concrete International, November 1991, pp. 56-64

6. Japan Society of Civil Engineering Association, Standard Specification for Design And Construction of Concrete Structures, 1996, Japanese version.

7. ASTM B 117-Reapproved 1979, Standard Test Method of Salt Spry (Fog) Testing, Annual Book of ASTM Standards", (Philadelphia, PA: ASTM).

8. ASTM C876-77, "Standard Test Method for Half Cell Potential of Reinforced Steel in Concrete, Annual Book of ASTM Standards, Vol. 04.02 (Philadelphia, PA: ASTM).

9. Report of the Technical Committee 60-CSC RILEM, Edited by P. Schiessl, "Corrosion of Steel in Concrete", 1988, p.76

10. COMITE EURO-INTERNATIONAL DU BETON, Durability of Concrete Structures-State of the Art Report, Bulletin O'Information No. 148, Paris, France, 1982.

11. AYALA, V, INGALLS, A AND GENESCA, J. Mathematical Modeling of Rebar Corrosion in Seawater, On-Line Conferencing, 1997 and M. A. Tullmin, C. M. Hansson and P. R. Roberge," Electrochemical Techniques For Measuring Reinforcing Steel Corrosion", 1997, http://www.clihouston.com/intercorr/techsess/papers.

12. WEST, Basic Corrosion and Oxidation, Published by Ellis Horwood Ltd, England, pp. 155, 1980.

13. ZHANG, X G, AND VALERIOTE, E M. Corros. Sci., 34, 1957 (1993).

CRITICAL LOCAL FAILURE OF EXTERNALLY STRENGTHENED CONCRETE ELEMENTS LOADED IN FLEXURE

V Bokan-Bosiljkov

V Bosiljkov

S Gostič

R Žarnić

University of Ljubljana

Slovenia

ABSTRACT. Short reinforced concrete beams with epoxy-bonded steel or carbon fibre reinforced plastic (CFRP) plates were tested in bending. Specimens were designed to explore critical situations when shear or peeling failure prevailed over flexural failure. The results of experimental testing showed the behaviour of the specimens was significantly influenced by the type of plates used. The analytically calculated local peeling stress at the end of the plate can indicate the observed failure when externally bonded plate separates from the concrete surface of the specimens. A 3D numerical modelling with homogenized approach indicates that further improvement of the model should be carried out, particularly where shear behaviour of the materials has a prevailing effect on the failure of the specimens.

Keywords: Bending strengthening, Carbon fibre reinforced plastic (CFRP) plates, Steel plates, Epoxy adhesive, Short span beams

Dr Violeta Bokan-Bosiljkov is a Teaching Assistant in Building Materials at the Faculty of Civil and Geodetic Engineering, University of Ljubljana, Slovenia. Her main research interests is concrete technology including modified concrete.

Vlatko Bosiljkov, MSc is Ph.D. student at the Faculty of Civil and Geodetic Engineering, University of Ljubljana, Slovenia. His main research interests are experimental investigation and numerical modelling of concrete and masonry elements.

Samo Gostič, MSc is Ph.D. student at the Faculty of Civil and Geodetic Engineering, University of Ljubljana, Slovenia. His main research interest is in development of experimentally supported computational models of structures.

Professor Roko Žarnić holds the Chair for Research in Materials and Structures, Faculty of Civil and Geodetic Engineering, University of Ljubljana, Slovenia. Lately, his main research has been oriented on strengthening of reinforced concrete structures.

INTRODUCTION

It is often necessary to strengthen an existing reinforced concrete (RC) structure during its life span. The causes for strengthening may be various, including reduced load-bearing capacity due to damages of RC elements and errors that may have been made during the design and/or construction phase of the structure. On the other hand, it is not unusual that demands regarding the serviceability of the structure change with time in such way that from then on the structure has to carry larger loads.

If insufficient reinforcement is the reason for any of previously mentioned situations, the improvement of the bending or shear capacity of concrete elements by means of externally bonded plates could be an efficient solution. Although the most common material used for strengthening is still steel, the application of carbon fibre-reinforced plastic (CFRP) is becoming more frequent. This is due to a number of difficulties connected with the process, based on steel plates external reinforcement [1]:

- steel oxidation: steel requires a careful protection and maintenance,

- the mechanical properties of steel can not be fully utilised,

- before bonding, a specific surface treatment is required,

- bonding of steel plates must be achieved under pressure,

- steel is heavy, very stiff and quite difficult to handle,

- this technology could not be applied on wide and curved surfaces.

The aim of our experimental work was to study the critical situation when peeling and or shear failure prevailed over the flexural failure of the concrete reinforced beams strengthened in flexure with externally bonded steel as well as CFRP plates. Also the influence of the external reinforcement to the global and local deformation ability of beams was studied.

EXPERIMENTAL PROGRAMME

Materials and Test Specimens

The test beams were 150x150 mm in cross-section and 1000 mm long. They were made by using concrete mix of strength class C30/37 (f_{ck}=30 N/mm^2). For the internal reinforcement ribbed bars $2\phi12$ (f_{yk}/f_{tk} = 400/500 N/mm^2) at an effective depth of 120 mm and mild steel stirrups $\phi6$ were used. Along the central segment of the beam (at the length of 500 mm) the spacing between the stirrups was 125 mm, while along the end segments of the beam (at the length of 250 mm) the spacing was 75 mm. For the external reinforcement, mild steel plates as well as CFRP plates were applied. The cross-section of the used steel plates was 4x60 mm^2 with ultimate mean tensile strength of 235 N/mm^2 and E-modulus of 210 kN/mm^2. The CFRP plates (SikaCarboDur) had cross-section equal to 1.2x40 mm^2, ultimate tensile strength of 2400 N/mm^2 and E-modulus of 150 kN/mm^2. The plates were 800 mm long (Figure 1). For gluing the plates to the concrete surface, the epoxy adhesive Sikadur-30 with compressive strength of 100 N/mm^2, bond tensile strength with concrete of 4 N/mm^2, bond shear strength with concrete of 15 N/mm^2 and E-modulus of 12,8 kN/mm^2 was applied.

Testing

Beams were tested under four-point bending. The bending span between the supports was equal to 900 mm and the points for load application were on the thirds of the span. For each type of the external reinforcement, two beams were tested together with one reference specimen without externally bonded plate. The beams were loaded by displacement controlled servo-hydraulic actuator INSTRON with capacity of 250kN. The loading speed in the elastic range was equal to 5kN/min. During each test flexural forces and deformation characteristics of the test specimen were simultaneously measured and crack pattern was traced and registered. Deflection of the beam was measured by LVDT transducer and concrete strains along the beam height were measured by electronic deformeters (Figure 1). Strain gauges were used for the measurements of the concrete as well as plate strains at the lower surface of the beam, as it is shown in Figure 1.

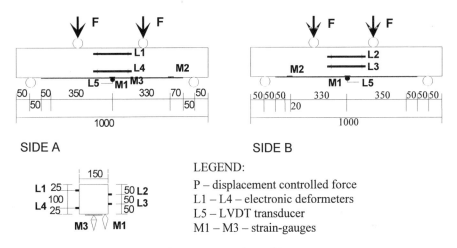

Figure 1 Layout of tested beams

TEST RESULTS AND DISCUSSION

The performed bending tests yielded a considerable amount of data on the influence of the externally bonded plates to the global as well as local behaviour of the short RC beam up to its post peak load-bearing capacity. In Figure 2 the flexural load-deflection curves for the reference beam, two test specimens strengthened with the steel plate and one beam strengthened with the CFRP plate are presented. Due to the measurement error, the obtained results for the beam deflection are not reliable in the case of the second beam with externally bonded CFRP plate. From Figure 2 we can see that the steel plate considerably increased the stiffness of the beam up to the first drop of the load. In case of the CFRP plate the increase in the beam stiffness was not significant. The multiple drop of the flexural load during the test is characteristic of the load-deflection curves representing the behaviour of the strengthened beams. The localisation of the shear-peeling crack at the plate end initiated the separation of the strengthening plate from the specimen, which is represented in Figure 2 as the first drop of the load. The following drops of the load resulted from the progress of the same process.

Since at the reference beam the yielding of the steel bars occurred before the ultimate load was applied, the beam failure was caused by flexural-shear failure mode. The experimentally obtained results were confirmed also by theoretical calculations of the ultimate shear and bending load-bearing capacity of the reference beam, which both gave a value approximately equal to 80 kN. On the other hand, externally bonded plates held the concrete together and delayed and hindered crack development. Therefore, yielding of the steel bars did not occur and the strengthened beams failed due to shear failure mode.

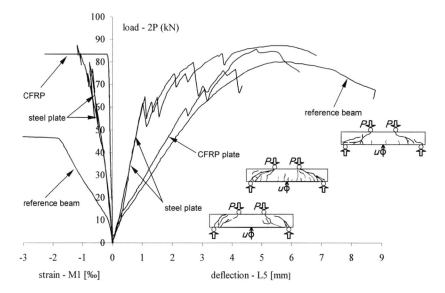

Figure 2 Flexural load-deflection and flexural load-concrete strain curves for the tested beams

Precisely, soon after the flexural cracks occurred in the middle span between the points of load application, flexural cracks at both ends of the plate were initiated. With increased loading this cracks transformed into diagonal tensile cracks, which were arrested by the internal stirrups. At the same time the separation of the strengthening plate occurred due to the localisation of a shear-peeling crack in concrete just above the glue layer. As the process of the plate separation continued, one of the diagonal tensile cracks started to open and crack faces started to slide mutually. Due to relatively small spacing between the stirrups, beside the localised diagonal tension crack also more distributed shear cracks occurred in the shear span. Similar failure mechanism was observed also by Jansze [2].

We can conclude from the measurements of the concrete strains along the beam height at the middle of the beam span (L1 – L4) that delaying and hindering the crack development by the strengthening plate resulted in considerable change of strain distribution along the cross-section. While strain diagrams obtained at the reference beam are actually straight lines up to the peak load, this was not observed for the strengthened beams. In the last case it seems that after the separation of the strengthening plate has began, the shape of the strain diagrams changed from straight line to bi-linear one.

In particular, along the lower 50 mm of the height of the cross-section the concrete strains seem to be approximately constant (steel plate) or they even declined towards the lower strains (CFRP plate).

ANALYTICAL APPROACH

The shear and peeling stresses at the end of the strengthening plate, in a moment when the plate separation started, were estimated. For this purpose the closed analytical formulae derived by Täljsten [3] were used. As the input point force for the formulae the experimentally obtained value at the first drop of the load (2x30 kN Figure 2) was used. The thickness of the adhesive layer equal to 1.5 mm for the steel plate and 0.5 mm for the CFRP plate was chosen.

The obtained results are presented in Figure 3. We can see that the calculated peak shear stresses are considerably lower than the bond shear strength between the epoxy layer and the concrete. On the other hand, peeling stresses are close to the bond tensile strength between the glue layer and the concrete. Therefore, it can be concluded that high peeling stresses at the end of the strengthening plate were responsible for the separation of the steel as well as CFRP plate from the beam surface. Due to good agreement between the experimentally and the analytically obtained results it seems that the shear and peeling stresses in the bond zone could be controlled by theoretical calculations. This approach was already used for the design of long bending RC elements with externally bonded steel and CFRP plates, which should increase the bending load-bearing capacity of the elements for at least 30% as compared to the reference.

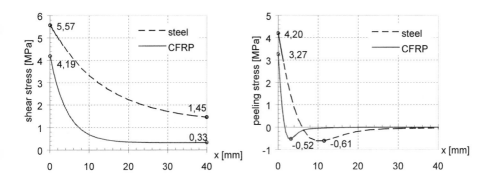

Figure 3 Shear and peeling stresses at the end of the strengthening plate in the bond zone

NUMERICAL MODELLING

In order to simulate the behaviour of flexurally loaded strengthened short span beams, a non-linear 3D modelling was performed. For each series of the reference and strengthened specimens two different types of numerical models were made: The first one - a homogenized model, where the RC beam was modelled as a plain material and the second one – a detailed model, where the concrete and the reinforcement within the beam were modelled separately.

All the material properties for the numerical models were used according to the already mentioned properties of materials and specimens. More detailed description of the used models is given in Žarnić [4].

With the homogenized models the overall behaviour of the short-span beams were much softer as it was the case for the detailed models. On the other hand, the detailed models had a almost the same initial stiffness as the experimental specimens (Figure 4 –(a)), but due to lack of the shear band within the non-linear program they could not successfully modelled the localisation of the shear-peeling crack at the plated end which is presented in Figure 4 –(b).

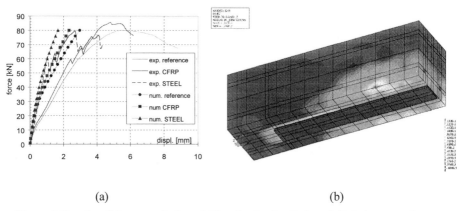

| (a) | (b) |

Figure 4   Results of the numerical modelling for the detailed modelled short span beams

CONCLUSIONS

1. With the use of steel plate for the bending strengthening of the RC beam, the stiffness of the beam considerably increases up to the initiation of the shear-peeling crack at the end of the plate.

    In the case of the CFRP plate the increase of the beam stiffness is not significant. On the other hand the flexural load at the moment of localising the shear-peeling crack at the plate end, which initiates the plate separation, was approximately the same for the steel as well as for the CFRP external reinforcement.

2. Furthermore it is obvious that the externally bonded plates hold the concrete together and delay and hinder crack development. This is probably the cause for significant change of strain distribution along the central cross-section of the beams, detected during the tests. After the initiation of the strengthening plate separation, the shape of the strain diagrams changed from straight to bi-linear line.

3. Due to good agreement between the experimentally and the analytically obtained results it seems that the shear and peeling stresses at the end of the strengthening plate (in the bond zone) could be controlled by theoretical formulae, derived by Täljsten [3].

REFERENCES

1. LUYCKX J. at all, Bridge strengthening by carbon fibres. The Concrete Way to Development, FIP symposium 97, Johannesburg, South Africa, Symposium papers, 1997, Vol. 2, pp. 867-872.

2. JANSZE W., UIJL J., WALRAVEN J., Shear capacity of beams with externally bonded steel plates. The Concrete Way to Development, FIP symposium 97, Johannesburg, South Africa, Symposium papers, 1997, Vol. 2, pp. 857-866.

3. TÄLJSTEN B., Strengthening of beams by plate bonding. Journal of Materials in Civil Engineering, November 1997, pp. 206-212.

4. ŽARNIĆ R., GOSTIČ S., BOSILJKOV V. AND BOKAN-BOSILJKOV V., Improvement of bending load-bearing capacity by externally bonded plates. Proc. of Int. Conf. on Creating With Concrete, Dundee, September 1999.

REPAIR OF DEFICIENT SHEAR STRENGTH BEAMS BY CFRP

T Javor

S Priganc

A Alarashi

Technical University of Košice

Slovakia

ABSTRACT. Reinforced concrete beams with external strengthening by carbon fibre reinforced plates CFRP were analysed. The beams with deficient shear strength were damaged to appearance of shear crack and then repaired by CFRP. The experimental results are illustrate in this paper. Experimental data on strength, tension and compress strain, deflection, and mode of failure of the beams were obtained, and comparison between the repaired and the unrepaired beams were obtained.

Keywords: Composite material, Bonding, External prestressing, Repairs, Carbon fibre, Shear strength.

T Javor DSc.is Professor of Concrete structures in the Technical University of Košice and Direktor of EXPERTCENTRUM Bratislava. He specialises in experimental analysis of concrete structures, mainly bridges. He is member of several scientific organisations and he has published over 35O papers and several books.

S Priganc is Assoc. Professor, is Head of Department of Concrete Structures and Bridges, Faculty of Civil Engineering, Technical University of Košice, Slovakia. His research interest include the use CFRP in rekonstruction of Civil Engineering Structures.

Mr A Alarashi is a PhD student at the Technical University of Košice. He is currently completing his PhD on the subject CFRP.

INTRODUCTION

A reinforced concrete beam must be designed to develop its full flexural strength to ensure a ductile flexural failure mode under extreme loading. Hence, a beam must have dangerous and less predictable than flexural failure. Shear failure of RC (diagonal tension failure) is a type of failure mode which has a catastrophic effect, should it occur. If a RC beam deficient in shear strength is over loaded, shear failure may occur suddenly with out advance warning of distress, while a flexural failure occurs gradually (if the beam is under reinforced), with large deflections and cracking giving ample warning. Therefore, a RC beam must have sufficient shear strength to ensure ductile flexural failure.

EXPERIMENTAL STUDY

The behaviour of RC beams deficient in shear capacity and strengthened with the carbon fibre reinforced plates (CFRP) was investigated by conducting flexural tests on these beams. Six reinforcement beams with a cross-section of 120x200 mm and length 1600 mm were tested. Tensile reinforcement was provided by three 12 mm diameter bars in the bottom, and two 6 mm diameter bars in the top. The web reinforcement consisted of 6 mm diameter closed stirrups, spaced 130 mm centre to centre throughout the span of the beam. Detail of the batch beams are shown in Figure 1.

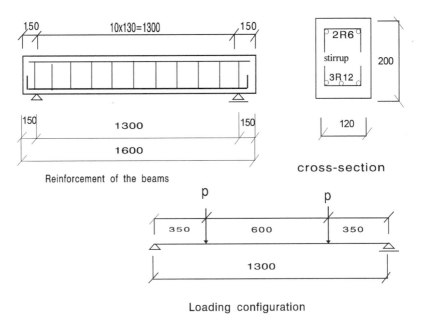

Figure1 Reinforcement of the beams and scheme of the loading configuration

The test is done in two stages: the first stage is to simulate the damage, by reloading under two incrementally symmetrical point loads (see Figure 1), up to the max. width shear cracks of about 0.02-0.025 mm. The load was then removed and beams were repaired by CFRP in tow alternative as shown in Figure 2. In the second stage the beams were loaded under two incrementally symmetrical point loads until failure. The beams were strengthened by Sikadure laminate CFRP, and the joint with concrete was fixed by the epoxy adhesive (Sikadure 30).

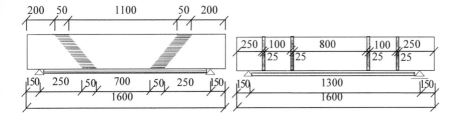

Figure 2 Alternative strengthening of the beams

The max. strain of the beams was measured by electrical resistance strain gauges mounted to the concrete surface and the laminate. The strain was recorded automatically by apparatus UPM60, which connect with PC.

EXPERIMENTAL RESULTS

An iterative analytical method has been used for predicting the strengthened beam response to load application. This analytical model uses the principles of strain compatibility and equilibrium, and the material constitutive relation for concrete, steel and laminate by theory elasticity, which can then be compared with experimental data.

The shear strain was obtained by :

$$\sigma_x = \frac{M}{w}$$
$$\sigma_y = 0$$
$$\tau_{xy} = \frac{P \cdot S}{I \cdot b}$$

where,

M is bending moment in the place of the gauge,
P is the load,
S is static moment to the centroid of the cross-section,
b is the width of the cross-section,
I is the moment of inertia.

Whilst the strains were obtained using the following equations:

$$\varepsilon_x = \frac{1}{E}\left(\sigma_x - \mu\,\sigma_y\right)$$

$$\varepsilon_y = \frac{1}{E}\left(\sigma_y - \mu\,\sigma_x\right)$$

$$\varepsilon_z = \frac{1}{E}\left[\frac{E}{2}\left(\varepsilon x + \varepsilon y\right) + \tau_{xy}\left(1 + \mu\right)\right]$$

The strain at the beams were measured in the place where is prediction the maximum shear as it shown in Figure 3. In Figure 4 shows the strain in the beams after loading and after repairing in the first alternative of repairing. In Figure 5 shows the strain in the beams after loading and after repairing in the first alternative of repairing.

CONCLUSIONS

CFRP laminates can provide increase in strength and stiffness to existing concrete beams when bonded to the web and tension face. The magnitude of the increase and the mode of failure are related to the direction of the reinforcing fibres. When the laminate were placed perpendicular to cracks in the beam (first alternative), an increase in stiffness and strength was obtained up to 38% and a brittle failure occurred due to concrete rupture as a result of stress concentration near the ends of the laminates. When the laminate were placed perpendicular to the steel bars (second alternative), an increase in stiffness and strength was obtained up to 40%. Strengthening damaged beams with composite laminate CFRP can be used to effectively increase the strength and ductility of deficient concrete beams.

REFERENCES

1. GARDEN H N et al. The strengthening and deformation behavior of reinforcement concrete beams upgraded using prestressed composite plates, Materials and Structures, Vol. 31, May 1998.

2. DEURING M et al.Verstärken Von Stahlbeton mit gespannten Faserverbund-werkstoffen, Dissertation EMPA-Bericht Nr.224, Dubendorf 1993.

3. ARDUINI M et al. Brittle Failure in FRP plates and sheet bonded Beams, ACI Structural Journal July-Aug., 1997.

4. SWAMI R et al. Repair and Strengthening of Concrete Members with Adhesive Bonded Plates, ACI Michigan, 1997

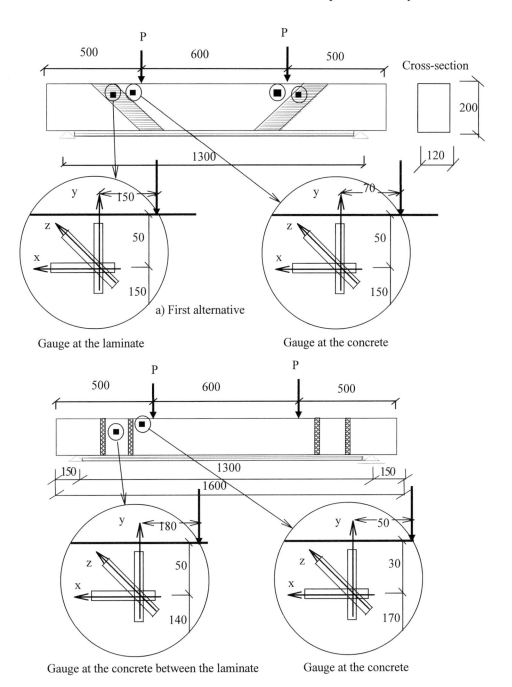

a) First alternative

Gauge at the laminate   Gauge at the concrete

Gauge at the concrete between the laminate   Gauge at the concrete

b) Second alternative

Figure 3 The places of the gauge strain after repairing

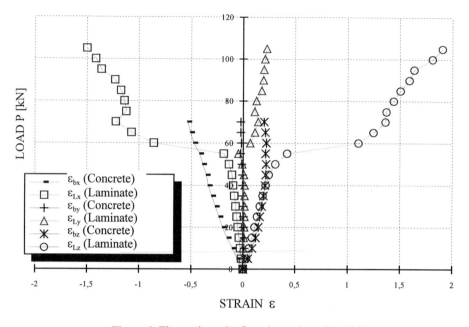

Figure 4 The strain at the first alternative of repairing

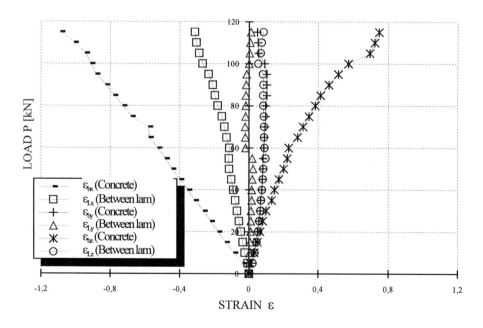

Figure 5 The strain at the second alternative of repairing

HIGH PERFORMANCE STRUCTURAL ELEMENTS USING CARBON FIBRE REINFORCED POLYMERS

N K Subedi

N R Coyle

University of Dundee

United Kingdom

ABSTRACT. Carbon Fibre Reinforced Polymer, CFRP, is being increasingly used in civil engineering structures. For strengthening of existing concrete structures, the material has been extensively tried in concrete beams for enhancement in flexure, and in shear. In masonry structures, attempts have been made to strengthen shear walls. CFRP is now being suggested as cables for cable stayed bridges. In this paper, the feasibility of utilising CFRP strips, as main reinforcement embedded in concrete beams is investigated. Two beams with encased CFRP strips in the form of I-section and truss were constructed and tested in flexure to failure. The results of the test are discussed. The test proves that CFRP embedded in concrete provides an efficient form of reinforcement, and with well-distributed and uniform cracks in the stressed area. The failure load obtained in one test was 41% higher than that predicted by strain-compatibility approach.

Keywords: Carbon fibre reinforced polymer (CFRP), Concrete, Composite beam, Flexure, Failure load, Strain-compatibility.

Dr Nutan Kumar Subedi is a senior lecturer in civil engineering at the University of Dundee, Scotland, UK. His main research interests are in the field, analysis and design of concrete structures, which include deep beams, panel structures, tall buildings, shear walls, concrete/CFRP and concrete/steel composite beams.

Neil R Coyle is a research student in the Department of Civil Engineering, University of Dundee. His research area is composite beams.

INTRODUCTION

Carbon Fibre Reinforced Polymer (CFRP) material is being increasingly used in civil engineering structures as strips or laminates, as fabric wraps and as cables in the form of parallel wire bundles. Research so far has mostly concentrated in its use in repair and in strengthening of existing structures. The range of applications cover concrete elements [1], masonry structures [2] and the possibility also exists for strengthening timber elements. The method of strengthening is typically by attaching the CFRP strips or laminates by means of specially formulated adhesives to the external surface of the structure. When set the CFRP laminates act compositely with the parent material resisting any applied load in a composite manner.

CFRP is a lightweight material with very high uni-axial tensile strength characteristics. Its lightweight characteristics make it easy to handle, work and fix on the structures. Currently the cost of the material is very high, but, with its potential for widespread application the process of manufacture could be improved and cost brought down to acceptable levels.

In this paper the use of CFRP as embedded reinforcement is examined. Two beam elements are investigated. In one case CFRP strips are encased in concrete section in the shape of I-section. This is called the I-configuration. In another case, the encased CFRP strips were in the form of a truss and thus the name truss configuration. The main objective of the study was to establish the feasibility of utilising CFRP as embedded reinforcement in concrete beams. The experimental results are presented and a comparison is made with the strain-compatibility method of analysis.

BEAM CONFIGURATION

For all the test described here Carbon Fibre Reinforced Polymer (CFRP) laminates 50 mm wide x 1.2 mm thick strips supplied by Sika Ltd were used. Two configurations of CFRP strips were considered for test. The first one is in the shape of I-section assembled from 50 mm strip laminates and held together by small lengths steel angle cleats 25 x 15 and strips 13 mm wide x 3 mm thick as shown in Figure 1 (a). A view of I-section configuration during construction is shown in Figure 1(b).

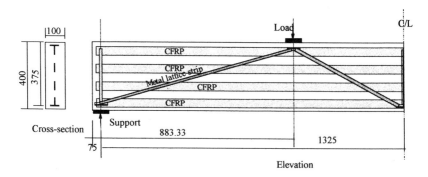

Figure 1(a) CFRP/Concrete beam: I-section configuration

Figure1(b) I-section configuration during construction

The second configuration consisted of a truss form as shown in Figure 2. In this specimen the bottom member consisted of two strips held together, but not glued, to increase the area of section and to provide approximately the same amount as the I-section.

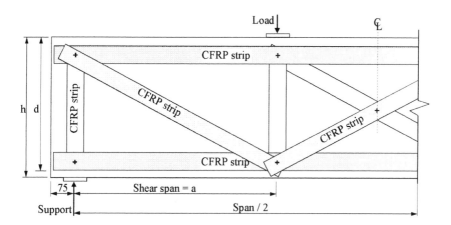

Figure 2 CFRP truss configuration

In both specimens, 25.4 mm square weld mesh was provided near the vertical surfaces for surface crack control purposes. The overall size of the beams was 100 wide x 400 deep x 2800mm long. The centre to centre distance between the supports was 2650 mm. The beams were designed to be loaded at third points creating the middle third free of shear forces.

Casting

The concrete strengths were, I-section: f_{cu} = 49.4 N/mm^2 and truss section: f_{cu} = 58.3 N/mm^2. Prior to casting the CFRP strips were treated with a coating of adhesive as a bond-bridging layer between the concrete and CFRP. The adhesive used was Sikadur 31, also supplied by Sika Ltd.

STRUCTURAL BEHAVIOUR

The test arrangement followed a standard set up with loading beam and simple supports allowing for translation and rotation. The test was carried out in at least three cycles of loading and unloading before taking the beam to failure. The test set up is shown in Figure 3.

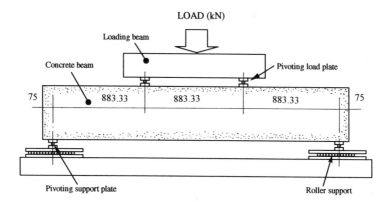

Figure 3   Test setup

Behaviour under load: I-section configuration

The I-section configuration, Figure 1, was tested in five cycles. In the first four cycles, the loads were taken up to the maximum values of 29.4, 58.9, 88.3 and 117.7 kN. In the fifth cycle the beam was loaded to 147.2 kN followed by each subsequent increment of 9.81 kN. The beam failed at 274.7 kN.

Throughout the test the development of cracks was uniform; starting in the central region of the beam and spreading throughout the soffit. Both types of cracks, flexure and flexural shear, were well formed. The uniformity of cracks throughout the length of the beam was a clear indication of the proper bond between the concrete and CFRP strips. The progress of cracks on the beam showed a gradual shift of neutral axis in the upper region. The crack widths measured at regular intervals gave a maximum crack width of 0.3 mm at about 90 kN load. At failure some cracks progressed up to approximately 80 mm from the top of the beam. Two stages of the formation of cracks are shown in Figure 4.

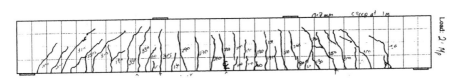

(a) Crack pattern at approximately 200 kN

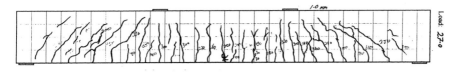

(b) Crack pattern at approximately 270 kN load

Figure 4 I-section configuration: crack patterns at higher loads

The failure was characterised by the formation of a substantial length of crushing of the top part of concrete in the central region. As soon as the crushing was completed, the load dropped and the cracks were closed indicating that the CFRP reinforcement remained elastic throughout. There was neither sign of horizontal cracks nor any other form of premature failure. Figure 5 shows substantial damage to the concrete in the extreme compression region, but all visible cracks have now closed.

Figure 5 View of I-section configuration after failure

Behaviour under load - Truss configuration

The truss configuration, Figure 2, completed its test in three cycles. As before, in the first two cycles the loads were taken up to 29.4 and 58.9 kN. In the third cycle, the load was gradually increased to 58.9 kN and then followed by increments of 9.81 kN until failure at 188.4 kN was recorded.

As in the I-configuration the development of cracks in this beam was uniform, at regular spacing and both flexure and shear cracks forming along the length of the beam. This continued to happen up to a load of 176.6 kN. At an attempt to increase the load at this moment, a crack right along the edge of one of the loading plates, Figure 6(a) increased in width and also progressed along the height. This was clearly the critical crack, which led to failure at the load of 188.4 kN. On examination, it was revealed that there was a longitudinal crack on the soffit of the beam, Figure 6(b), extending from the support to the position of the critical flexural crack. The formation of this crack is considered to be the cause for a lower failure load for this beam.

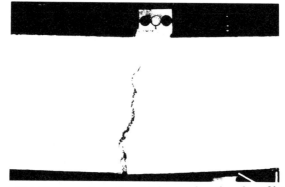

Figure 6(a) Truss configuration: Major crack at the edge of loading plate

Figure 6(b) Truss configuration: Horizontal split along the soffit of beam

DISCUSSION OF RESULTS

Load-Deflection Characteristics

A load-displacement plot of the two configurations for the final cycle of loading is shown in Figure 7. The plot shows deflections for each case at two different positions, under the load and at mid-span. As expected, in each case, the deflection at mid-span is slightly higher than under the load. The graph indicates that for both configurations, the load deflection relationship is linear for most part of the loading. The linearity is more pronounced in the case of I-configuration right up to failure. This indicates that the process of crack formation is more of a gradual process with no sudden deterioration in stiffness. It is also the characteristic of a well-distributed crack pattern. For these specimens the I-section shows a stiffer section than the truss configuration.

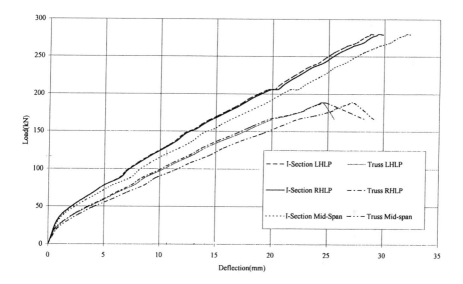

Figure 7 Load-deflection characteristics for the beams

The maximum deflections for the I-configuration are approximately 32.5 mm and 30 mm at the mid-span and load positions respectively. A deflection of 10.6 mm (span/250) representing the serviceability limits would give a serviceability load of approximately 115 kN representing 42% of the failure load.

The failure of the truss configuration at 188.4 kN is marked by a peak load followed by a drop. The peak deflections are approximately 27.5 mm and 25 mm at the mid-span and at the load positions respectively. Using deflection limits, the serviceability load for the truss configuration is approximately 95 kN representing 50% of the failure load.

Mode of Failure and Ultimate Strength

The mode of failure in the case of I-configuration is flexure with the crushing of most part of concrete in the high compression area. A strain-compatibility analysis of the mid-span section of the beam was carried out, Figure 8. The calculated ultimate load was 195 kN compared to the failure load of 274.7 kN, an increase of 41 %. The enhancement in failure load is attributed to the well-formed and well-distributed cracks in the beam as opposed to a crack concentrated at one particular location. The enhancement in load is a positive advantage of CFRP system and it requires further study to determine how exactly this is realised in beams.

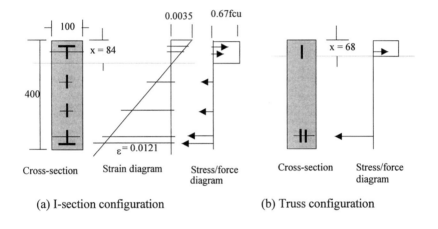

(a) I-section configuration   (b) Truss configuration

Figure 8 Strain-compatibility analysis

The well-distributed cracks also resulted in lowering of the strain at mid-span section, see Figure 9. The maximum strain in the CFRP strip at the mid span section is approximately 9900 micro-strains (extrapolated), which is less than the strain-compatibility prediction (Figure 8a) of 12100 micro-strains. Both the calculations and the experimental observation show that the CFRP strips are well under the maximum strains for failure. There are two conclusions that can be drawn from these observations (a) that only a small amount of CFRP will be sufficient for resisting loads in normal situations (b) the beams will act more efficiently with well distributed cracks.

The failure of the truss configuration was sudden splitting of the beam horizontally along the soffit followed by a sudden widening of the vertical crack along one edge of the loading plate. The strain compatibility analysis for the truss configuration (Figure 8b) showed a failure load of 193 kN compared to its actual failure load of 188.4 kN. The premature nature of the failure meant that there was no clear evidence of compression crushing apparent and there was no enhancement of load at failure. Early stages of the test showed similar behaviour to that of the I-configuration and the conclusions apply to both types of design.

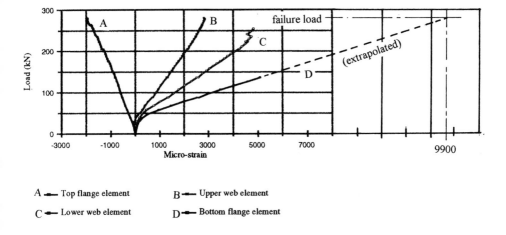

A —■— Top flange element   B —■— Upper web element

C —■— Lower web element   D —■— Bottom flange element

Figure 9 I-section configuration: Measured strains in CFRP at mid-span section

Compared to a conventionally singly reinforced beam with high-yield bars the failure loads represent as follows:

Table 1 Comparison between CFRP and conventionally reinforced beam

SECTION	FAILURE LOAD (kN)	x (mm)	CONVENTIONAL SINGLY REINFORCED BEAM, % REINFORCEMENT
I Section	274.7	116	2.71
Truss	188.4	62	1.40

CONCLUSIONS

The results presented in this paper suggests that:

1. Carbon Fibre Reinforced Polymer, CFRP, strips can be used as normal reinforcement in concrete beams.

2. With proper bonding, the composite section behaves very well in flexure.

3. Uniform and well distributed crack patterns are the hall marks of this type of construction, which help to reduce the longitudinal strains in the CFRP strips and increase the overall moment of resistance.

4. In the I-configuration embedded reinforcement; the experimental failure load was found to be 41% more than that predicted by the strain-compatibility method. For the truss configuration, although premature failure was suspected, the failure load compared well with the strain compatibility prediction.

5. Although, the enhancement of load is attributed to the proper bond and composite action between the concrete and CFRP, this aspect requires further study to establish the extent of the enhancement possible in such construction. This is a positive aspect of CFRP/concrete composite behaviour.

6. This preliminary work proves that CFRP used as embedded reinforcement in concrete beams works very well. The behaviour of the beams shows considerable promise for CFRP as internal reinforcing material in concrete structures.

REFERENCES

1. MEIRS, U.   Repairs using Advanced Composites, Composite Construction - Conventional and Innovative, Conference Report: Sept 16-18, 1997, International Conference Innsbruck 1997, pp113-124.

2. SCHWEGLER, G . Earthquake Resistance of Masonry Structures Strengthened with CFRP-sheets, Composite Construction - Conventional and Innovative, Conference Report: Sept 16-18, 1997, International Conference Innsbruck 1997, pp735-740.

ANALYSIS OF A WEDGE-TYPE ANCHORAGE SYSTEM FOR CFRP PRESTRESSING TENDONS

T I Campbell

J P Keatley

K M Barnes

Queen's University

Canada

ABSTRACT. Results from a finite element analysis of a wedge-type anchorage for a CFRP (Leadline) prestressing tendon are presented. It is shown that specific features of this anchor, namely the angular differential between the taper of the wedge and that of the inner surface of the barrel, and the coefficient of friction at the barrel/wedge interface, have significant influences on the stress patterns due to tendon loading on the anchor, particularly the radial stress on the tendon. Results from load tests simulating setting of the wedge in the anchorage are found to be in reasonable agreement with model predictions. It is concluded that the model is capable of simulating the general behaviour of the anchor but further development is necessary.

Keywords: Prestressed concrete, Anchorages, Carbon fibre reinforced plastic (CFRP) tendons, Finite element analysis, Anchor-set tests.

Dr T I Campbell is a Professor of Civil Engineering, Queen's University, Kingston, Canada. His main research interests relate to prestressed concrete transportation structures such as highway bridges and transit guideways. He is a project leader in ISIS Canada, a Network of Centres of Excellence on Intelligent Sensing for Innovative Structures, where his area of interest is corrosion-free, unbonded prestressed concrete systems.

Mr J P Keatley is a graduate student in the Department of Civil Engineering, Queen's University, Kingston, Canada. He headed the Queen's University team in the 1998 Canadian National Concrete Canoe Competition.

Ms K M Barnes is a former undergraduate student in the Department of Civil Engineering, Queen's University, Kingston, Canada.

INTRODUCTION

Steel tendons, which are used commonly to prestress concrete, are susceptible to corrosion, and as a result there is an interest in utilizing alternative non-corrosive materials, such as fibre reinforced plastics (FRPs), for this purpose. If tendons of such materials are used, particularly in post-tensioning applications, it is necessary to have anchorage systems which are also non-corrosive. Various forms of such anchors have been suggested by the manufacturers of FRP tendons, and can be classified as either bonded-type or wedge-type anchors, or a combination of both types [1, 2].

In a bonded-type anchor the tendon is bonded to the anchorage component, requiring the total anchorage system to be in place on the tendon prior to stressing. A wedge-type anchorage on the other hand can be mobilized on the tendon during the stressing operation resulting in a more flexible system than with a bonded-type anchor. However, particular problems arise with wedge-type anchors applied to carbon fibre reinforced plastic (CFRP) tendons.

CFRP tendons are strong in the axial direction but relatively weak in the transverse direction. This necessitates a lower radial pressure on the tendon due to wedging action than that utilized in a typical steel tendon anchorage. Further, since CFRP is a brittle material stress concentrations must be avoided when gripping the tendon and thus the use of teeth on the wedge, as is the case for steel tendons, is not desirable.

An analysis of a wedge-type anchor developed at the University of Calgary [3], specifically for 'Leadline' tendons [4], is outlined in this paper. The stress distribution and displacement in the anchor subjected to a loading condition representing that from a stressed tendon are described. The influences of the level of friction at the barrel/wedge interface, and the differential in the angles of taper between the barrel and wedge on the stress distribution and displacement are discussed.

Finally, modeling and load testing of the anchor-set behaviour are outlined, and calibration of the model with respect to anchor-set loading is discussed.

DESCRIPTION OF THE ANCHOR

The anchorage is shown diagrammatically in Figure 1. It consists essentially of a barrel containing a tapered four-piece wedge which grips the tendon. At present both the barrel and the wedge are made from stainless steel, but the use of other materials is being investigated [2]. A feature of the anchor is the difference in the angle of taper of the barrel to that of the wedge, namely 1.99° as opposed to 2.09°.

An anchor described by Kerstens et al. [5] has a similar feature. This feature implies that, prior to anchoring the tendon, contact between the barrel and the wedge will be made only at the top end (as oriented in Figure 1) of the anchor, and with increasing load in the tendon, contact will be developed over an increasing length of the barrel/wedge interface. A sleeve made from copper is used between the wedge and the tendon to minimize stress concentrations on the tendon.

ANALYSIS OF THE ANCHOR UNDER TENDON LOAD

The anchor was analyzed using the axisymmetric finite element analysis capability of ANSYS Version 5.3 developed by Swanson Analysis Systems [6]. PLANE42, a quadrilateral element having 8 degrees of freedom (2 per node) was utilized in modeling the barrel, wedge and tendon, while CONTAC12 (a node-to-node contact element) and CONTAC48 (a general surface-to-surface contact element) were used along the wedge/tendon and barrel/wedge interfaces, respectively.   Additional details of the model have been given by Campbell et al. [7].

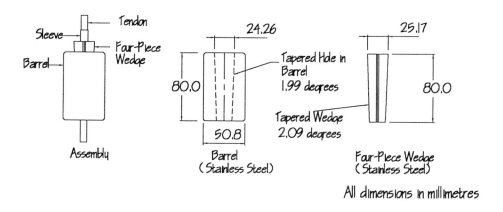

Figure 1 Details of the anchor

The barrel and the wedge comprised chromium-nickel stainless steel having a yield stress of 1230 MPa (N/mm²), while the tendon was assumed to be 8 mm diameter Leadline with an ultimate strength of 2550 MPa.   The barrel was modeled as a linear elastic isotropic material, while the wedges and tendon were modeled as linear orthotropic materials.   The four-piece wedge was simulated by specifying a low material stiffness in the circumferential direction in order to accommodate the axisymmetric analysis. The thin-walled copper sleeve was ignored in the model.

The material properties used in the model for the components of the anchorage are listed in Table 1 according to ANSYS notation.

In Table 1 MU is the coefficient of friction, KN is the stiffness of the contact element, EX, EY and EZ are the moduli of elasticity in the radial, (X), axial, (Y) and circumferential (Z) directions respectively, PRXY, PRYZ and PRZX are the major Poisson's ratios in the XY, YZ and ZX planes respectively, and GXY, GYZ and GZX are the corresponding shear moduli. A modulus of elasticity in the circumferential direction (EZ) of 0.1 MPa, together with zero values of the Poisson's ratios in the YZ and ZX planes, was used in simulating the four-piece wedge.

An axial load of 100 kN, corresponding to a stress of approximately 2000 MPa, was applied to the lower end of the tendon by means of an appropriate displacement which was increased in eight or ten sub-steps. The barrel was constrained against movement in the axial direction at its lower edge, and lateral movement of the tendon was prevented along its centreline.

Table 1 Material Properties of Anchor Assembly

MATERIAL PROPERTY	BARREL	WEDGE	TENDON	CONTAC 48	CONTAC1 2
Type	Isotropic	Orthotropic	Orthotropic	Contact	Contact
MU				0.2	0.4
KN (N/mm)				2,000,000	14,700,000
EX (GPa)	200	200	10.3		
EY (GPa)	200	200	147		
EZ (GPa)	200	0.0001	10.3		
PRXY	0.3	0.3	0.02		
PRYZ	0.3	0	0.27		
PRZX	0.3	0	0.27		
GXY (GPa)	77	77	7.2		
GYZ (GPa)	77	77	7.2		
GZX(GPa)	77	77	7.2		

Analyses were carried out for the anchor as detailed in Figure 1, and also for modified versions of the anchor in order to examine effects of change in the taper angle differential.

RESULTS FROM ANALYSES

Existing Anchor

The radial and circumferential stresses at the barrel/wedge and wedge/tendon interfaces, due to a load of 100 kN in the tendon, are shown in Figures 2 and 3, respectively.

Figure 2 shows that the radial compressive stress at the barrel/wedge interface increases rapidly over the upper 15 mm of the barrel and approaches a maximum at the top edge. Determination of the absolute maximum radial stress is difficult due to the finite size of the elements used in the mesh. Extrapolating from the trend of the radial stress distribution gives a radial stress at the upper edge of the barrel in the region of 330 MPa.

Figure 3 shows that the radial stress along the tendon/wedge interface increases from zero at both the top and bottom of the wedge to a maximum of 360 MPa at a location close to the top of the barrel.

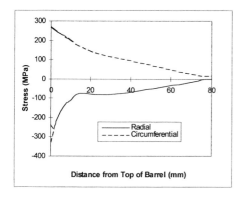

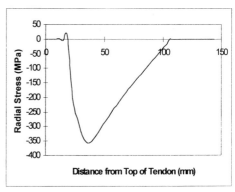

Figure 2   Radial and circumferential stresses
on barrel/wedge interface

Figure 3   Radial stress on wedge/
tendon interface

It appears that the ANSYS model gives a realistic prediction of the stress distribution in the anchor subjected to loading from the stressed tendon. Validation of the model is discussed later in this paper.

Effect of Friction at the Barrel/Wedge Interface

Figure 4 shows the radial compressive stress (indicated as negative) at the wedge/tendon interface, and the longitudinal tensile stress (indicated as positive) in the tendon for coefficients of friction of 0.05, 0.1, 0.2 and 0.3 in the barrel/wedge interface. The coefficient of friction at the wedge/tendon interface was assumed to be 0.4 in all cases. It can be seen that the magnitude of the radial stress on the tendon increases as the coefficient of friction decreases. This is to be expected since a reduced frictional coefficient in the barrel/wedge interface will require a larger radial force to resist the same force in the tendon.

Figure 4 also shows that, with the coefficient of friction of 0.05 which may be developed in a greased steel on steel interface, there is a significant radial stress gradient in the tendon at the bottom of the wedge. Such a stress gradient is undesirable since it may lead to brittle fracture of the tendon. Further, the longitudinal stress in the tendon is developed over a shorter length of the tendon as the coefficient of friction at the barrel/wedge interface decreases. In addition, it was found that the displacement required at the lower end of the tendon to induce a stress of 2000 MPa increased from 5.1 mm to 9.3 mm as the coefficient of friction decreased from 0.3 to 0.05.

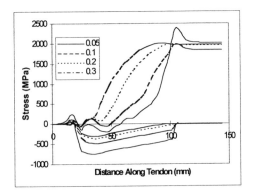

Figure 4 Variation of radial stress at the wedge/tendon
interface and stress in tendon with friction

It may be concluded that the level of friction in the barrel/wedge interface has a significant influence on the behaviour of the anchor. A low coefficient of friction can result in a stress gradient in the tendon at the lower end of the wedge and also in a larger anchor set.

Effect of Taper Angle

Figure 5 shows a comparison of the radial stress along the wedge/tendon interface for angular differentials of 0°, 0.06°, 0.11° and 0.2° between the taper of the wedge and the inner face of the barrel. In each case the angle of taper of the wedge was maintained constant at 1.95° while the angle at the inner face of the barrel was varied to give the desired differential. The coefficients of friction were 0.2 and 0.4 at the barrel/wedge and wedge/tendon interfaces, respectively.

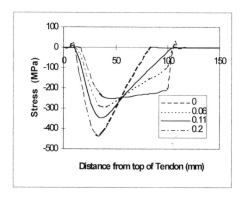

Figure 5 Variation of radial stress at wedge/ tendon
interface with angular differential

It can be seen in Figure 5 that the angular differential has a significant effect on the stress distribution. A 0° differential results in a compressive radial stress of approximately 220 MPa being developed in the tendon close to the bottom of the wedge, while a 0.2° differential produces zero radial stress in the tendon towards the bottom of the wedge, indicating that the wedge does not contact the barrel in this region.

On the other hand, the radial stress decreases to zero at the bottom of the wedge when the angular differential is 0.11°. Such a stress distribution is a desirable feature in that it avoids a sudden stress change, and the possibility of brittle fracture, in the tendon at the bottom of the wedge. It appears that an angular differential of around 0.1° is desirable with a coefficient of friction of 0.2 at the barrel/wedge interface.

ANALYSIS OF ANCHOR-SET LOADING

In order to grip the tendon in a wedge-type anchor, it is necessary to set the wedge by applying a load to compress it into the barrel in order to generate a compressive stress on the tendon, thereby inducing a frictional force in the tendon/wedge interface.

The anchor-set procedure was analysed using an ANSYS model similar to that developed for the anchor subjected to axial load in the tendon. With the barrel restrained in the axial direction, a displacement of the wedges into the barrel, by an amount requiring a load of 65 kN on the wedge, was induced. Analyses were carried out assuming coefficients of friction between the barrel and the wedge of 0.05, 0.1, 0.2 and 0.3.

Load-displacement results from the ANSYS model are presented in Figure 6. These results show a consistent pattern of stiffening of the anchor system with increasing load on the wedge, and that, at a particular load level, the displacement increases with a decreasing level of friction in the barrel/wedge interface.

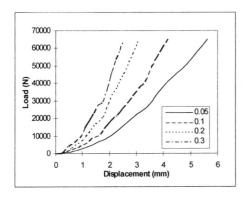

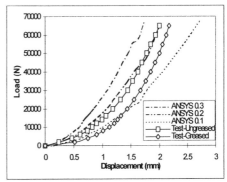

Figure 6 Load vs. displacement from ANSYS
model and tests for steel centrepiece

Figure 7   Load vs. displacement from
ANSYS model for varying friction

Load tests to simulate the anchor setting procedure were also performed in the laboratory. A load, which was increased up to 65 kN, was applied to the wedge, and the displacement of the wedge relative to the barrel at different load levels was measured. The 8 mm diameter Leadline tendon and copper sleeve, which form the centrepiece (portion gripped by the wedges) of the actual anchorage system, were replaced by a steel rod having the same diameter as the tendon-copper sleeve arrangement. In this way the influences of the copper sleeve and the uncertain orthotropic properties of the Leadline were eliminated, allowing direct comparison between test and ANSYS results.

Results from the two tests, one of which contained grease in the barrel/wedge interface, along with the ANSYS model predictions, assuming different values of the coefficient of friction in the barrel/wedge interface, are shown in Figure 7.

It can be seen from the test results that introduction of grease in the barrel/wedge interface increased the displacement of the wedge at a particular load level. Results from the test in which no grease was used are similar to those from the ANSYS model for a coefficient of friction in the barrel/wedge interface of 0.2, while those from the greased test are intermediate those for friction coefficients of 0.1 to 0.2. A coefficient of friction of 0.2 is not unrealistic for ungreased smooth steel surfaces, as is the case in the anchor, while a lower coefficient is expected for a greased steel surface. It may be concluded that the ANSYS model gives a realistic prediction of the anchorage behaviour in this case, and that the coefficient in the ungreased barrel/wedge interface is around 0.2.

A further two load tests were carried out in which the steel rod was modified by reducing its diameter to match that of the CFRP tendon, and fitting it with a copper sleeve. Grease was not used at the barrel/wedge interface in either of these tests.

The load-displacement behaviour, designated as ungreased steel-copper, observed in each of these two tests is shown in Figure 8, together with that shown in Figure 7 for each of the two tests with the steel centrepiece (labeled greased steel and ungreased steel).

Results from the tests with the steel-copper centrepiece show a pattern of behaviour having an initial large rate of displacement with load increase up to about 10kN, followed by a rapid increase in load with displacement at a rate similar to that observed in the tests with the steel centrepiece, but at an offset of around 1.7mm. Since the only difference between the two sets of tests was the presence of the copper sleeve in one set, the difference in behaviour may be attributed to deformation of the copper sleeve.

It appears that the copper sleeve yields at a load level of around 10kN, subsequent to which it is confined in the wedge/tendon interface, and serves only as a medium to transfer radial stress between the wedge and the tendon.

This observation justifies the neglect of the copper sleeve in the finite element analysis, however, while the stress distribution may be realistic, the displacement of the wedge relative to the barrel will be underestimated.

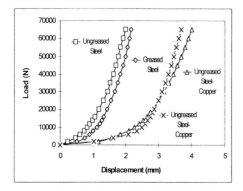

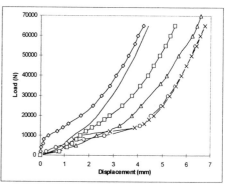

Figure 8   Load vs. displacement with steel   Figure 9   Load vs. displacement with CFRP-
        and steel-copper centrepieces   copper centrepiece

Six load tests incorporating CFRP-copper centrepieces in the anchor were carried out. Results are shown in Figure 9, where significant scatter is evident. Much of the scatter may be attributed to variation in the tightness of initial fit of the copper sleeve on the CFRP tendon. Due to the small angle of taper (2°) of the barrel, displacement of the wedge will be very sensitive to any lack of fit of the copper tube.

It may be concluded the finite element model appears to give a reasonable prediction of the behaviour of the anchor, with the exception that the displacement of the wedge relative to the barrel is underestimated due to neglect of the copper sleeve in the analysis.

CONCLUSIONS AND RECOMMENDATIONS

Based on results from this study the following conclusions and/or recommendations may be made:

1. The anchorage can be modeled using the axisymmetric finite element facility in ANSYS incorporating orthotropic material behaviour for relevant components and contact elements to simulate behaviour at interfaces between components.

2. A reduction in the coefficient of friction in the barrel/wedge interface results in an increased radial stress on the tendon. A significant stress gradient is induced in the tendon at the bottom of the wedge for a coefficient of friction of 0.05.

3. An angular differential of around 0.1° between the tapers of the wedge and the inner surface of the barrel, together with a coefficient of friction of 0.2 in the barrel/wedge interface, gives a radial stress of around zero in the tendon at the bottom of the wedge for a load of 100 kN in the tendon.

4. Displacement of the wedge relative to the barrel is under-estimated by the model as a result of the neglect of the copper sleeve.

5. The coefficient of friction in the ungreased barrel/wedge interface appears to be around 0.2.

ACKNOWLEDGEMENT

The authors gratefully acknowledge the financial support provided by the ISIS Canada Network of Centres of Excellence.

REFERENCES

1. NANNI, A, BAKIS, C E, O'NEIL, E F. AND DIXON, T O. Performance of FRP tendon - Anchor systems for prestressed concrete structures. PCI JOURNAL, Vol. 41, No. 1, 1996. pp 34-44.

2. REDA, M M, SAYED-AHMED, E Y AND SHRIVE, N G. Advanced composite materials for post-tensioning applications: Merits, shortcomings and possibilities. International Conference on Engineering Materials, Ottawa, Canada, pp 271-279.

3. SAYED-AHMED, E Y AND SHRIVE, N G. A new steel anchorage system for post-tensioning applications using carbon fibre reinforced plastic tendons. Can. J. Civ. Eng. Vol. 25, No. 1, 1998. pp 113-127.

4. MITSUBISHI CHEMICAL CORPORATION. Leadline brochure.

5. KERSTENS, J G H, BENNENK, W AND CAMP, J W. Prestressing with carbon composite rods: A numerical method for developing reusable prestressing systems. ACI Structural Journal, Vol. 95, No.1, 1998. pp 43-50.

6. SWANSON ANALYSIS SYSTEMS. ANSYS User's Manual for Revision 5.1.

7. CAMPBELL, T I, KEATLEY, J P AND BARNES, K M. Analysis of the Calgary anchor. Research Report, Department of Civil Engineering, Queen's University, Kingston, Canada, 1997. 33 pp.

THE MICROSCOPIC ANCHORING MECHANISM OF A FLUSH ANCHOR IN CONCRETE

C H Surberg

Hilti Corporate Research

Principality of Liechtenstein

B Meng

German Cement Works Association

Germany

ABSTRACT. The microscopic anchoring mechanism of a flush anchor was investigated. Optical interface analysis and quantitative image analysis were used to examine the interfacial interaction between anchor material and concrete as well as the pore size distribution in the concrete surrounding the anchor. Additional FEM-simulations of an expanded anchor were carried out and compared with the results from image analysis.

The results show that micro-keying with the aggregates within the concrete is a additional anchoring mechanism for this type of anchor. The results of the optical analysis correspond to the FEM-simulation of deformation distributions within the concrete.

Keywords: Anchor, Fastening technique, Concrete, Image analysis, FEM-Simulation

Mr C H Surberg received his master of science degree from the the Mineralogy-Petrographic Department of the University of Cologne (Germany). From 1992 to 1995 he worked at the Research Institute of the German Refractory Industry. There he was the head of the Mineralogy Laboratory. Since October 1995 he is responsible for the Construction Materials Research within the Hilti Corporate Research of the worldwide leading company in demolition tools and fastening technique in the Principality of Liechtenstein.

Dr B Meng studied Mineralogy at the University of Bonn (Germany). From 1986 to 1998 she worked at the Institute for Building Materials Research of the Aachen University of Technology (ibac). There she jointed the division „Binders and Concrete" and she was head of the labratory for Mineralogical and Microstructural Analysis. She finished her phD thesis on the subject of interrelationships between microstructural and moisture transport in 1993. In April 1998 B. Meng changed to the Research Institute of the Cement Industry, an Institution of the German Cement Works Association (VDZ).

INTRODUCTION

Fastening elements such as anchors are widely used for locally transferring loads into concrete structural components. Developing new and innovative products in fastening technology requires an in-depth understanding of the physical processes that form the basis of the anchoring mechanisms.

There are many investigations relating to the failure modes of a set anchor [1] [2] and relevant loading cases [3] (i.e. the setting and loading of the anchor). For the flush anchor, especially, the setting energy [4] and the load carrying capacity of anchors in cracked and uncracked concrete plays a central role [5] [6]. Numerical simulations of the interaction between the expansion sleeves and the concrete give further information about the anchoring [7] [8] [9]. They are based on the results of macroscopic fracture examinations in concrete [10] [11] [12]. There are some experimental investigation methods available to determine fracture mechanisms in concrete [13] [14] [15]. However, no experimental analysis about these sleeve-concrete interaction exists.

Hence, for a better understanding of the anchoring mechanisms affected by these interactions, we used a microscopic structure analysis technique combined with finite element simulation tools as a new approach. This paper presents the results of the experimental investigations and the numerical simulations of the concrete response for a completely expanded and unloaded HKD flush anchor.

THE HKD FLUSH ANCHOR

The HKD is designed for fastenings with small anchoring depth, i.e. channels/tracks, slabs/panels, brackets and mechanical installation components. It belongs to the displacement controlled expansion anchors and consists of a cone including a special cone safety device and an expansion sleeve (Figure 1).

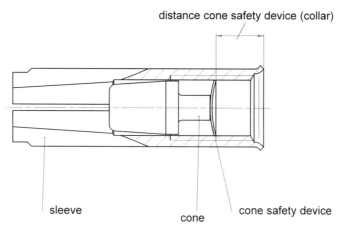

Figure 1   Technical design description of the Hilti HKD flush anchor

The setting procedure of the anchor is performed by hammering the cone until the end of the expansion sleeve reaches the bottom of the borehole (Figure 2 and Figure 3). This is done utilizing a special setting tool. The load is mainly transferred to the base material by friction.

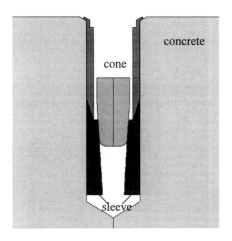

Figure 2 Axisymmetrical modelling of the HKD flush anchor

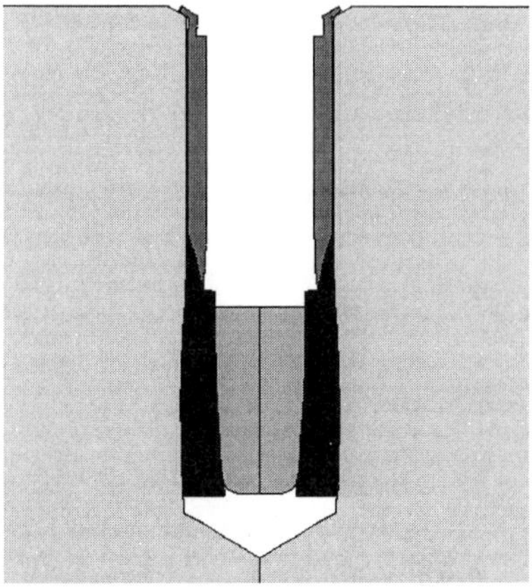

Figure 3 Displacement at the end of the setting procedure

The typical load-displacement behaviour of this type of anchor in two different concretes is shown in Figure 4. For detailed information on this product refer to [16].

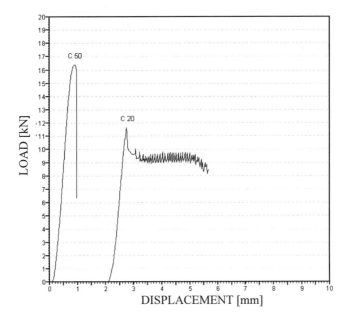

Figure 4   Characteristic pull out behaviour of HKD flush anchor in two different concretes

EXPERIMENTAL

The HKD flush anchors were set in concrete C20 with a maximum aggregate size of 4 mm. The anchors had an inner thread diameter of 6 mm (M6) and an outer diameter of 8 mm. The anchor hole was drilled using a rotary hammer and a carbide tipped drill bit. The depth of hole required for the used anchor was 25 mm. The bore hole was cleaned carefully to remove dust and fragments using a jet of air in order to guarantee perfect functioning of the anchor. Afterwards the anchor was set flush with the work surface.

In order to determine the interaction between the sleeve and the concrete components the microstructure was observed by transmitted light microscopy. Figure 5 shows a transparent microsection overview of an anchor set in concrete. It should be mentioned that the rigidity of concrete aggregates play a major role in the interaction of the sleeve with the surrounding concrete. Therefor the rigidity of aggregates was determined by microscopic indentation using the universal hardness test.

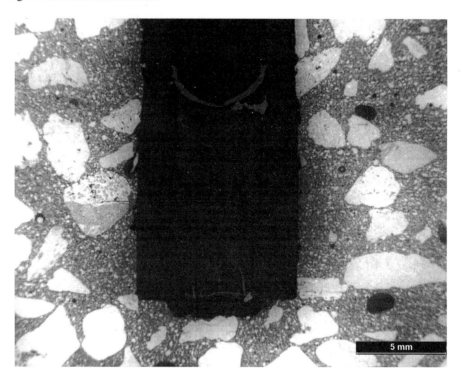

Figure 5   Overview of a HKD flush anchor set in concrete (optical micrograph)

Quantitative image analysis of optical micrographs were also used to determine the plastic deformation of the concrete microstructure. By means of the micropore size distribution the state of deformation was calculated. The deformation is only visible optically as a darkening of the cement matrix due to a reduction of capillary pores. The concrete used in this investigation had a capillary pore volume of 19 vol.% in the undeformed state. These were filled with blue coloured resin before the analysis. The blue saturation level of the optical signal was used as a parameter indicating the state of concrete compression and therefore representing the level of concrete deformation. For the measurement of the pore size distribution we used exclusively the blue channel signal of a RGB-CCD-camera. Afterwards using the different blue saturation levels of the capillary pores the phases were classified into three groups: low, middle and high porosity zones. To illustrate this differentiation we used a false grey scale composite.

For a cross-check of these experimental results the HKD flush anchor was modelled axisymmetrically in a completely expanded and unloaded case by a simulation tool developed by Hilti [8]. The geometry of the anchor and a part of the base material is shown in Figure 2 as well as the displacement at the end of the setting procedure (Figure 3). For the numerical calculation we used anchor dimensions and concrete input parameters which corresponded to the image analysis investigations.

The constitutive modelling of the concrete material was done using a uniaxial model. In the tension domain this model is based on the smeared crack approach [11]. The stress/strain relation is completed in the compression domain with a nonlinear hardening function up to the compressive strength of the concrete material.

RESULTS AND DISCUSSION

In the case of the HKD flush anchor macroscopically the tensile load is transferred to the base material mainly by friction. In addition to friction aggregates indent the expansion sleeve. This leads to a micro-keying effect increasing the load bearing capacity. As we can observe in Figure 6 and Figure 7 polycrystalline quartz-aggregates penetrate the expansion sleeve at the bottom and at the side of the sleeve with varying depth of indentation.

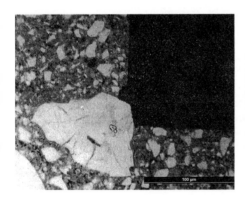

Figure 6 Micro-keying effect at the bottom of the expansion sleeve: penetration of a poly- crystalline quartz-aggregate (optical micro- graph)

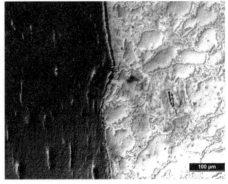

Figure 7 Micro-keying effect at the margin of the expansion sleeve: surface scratching of a polycrystalline quartz-aggregate (optical micrograph)

However, this micro-keying effect takes place only at a microscopic level and was only observed for aggreagtes with high rigidity, i.e quartz. Table 1 shows the results of a microscopic indentation test of two different aggregate types describing their rigidity behaviour. Due to the low Vickers hardness (HU_{plast}) and universal hardness (HU) and the low derived modulus of elasticity (E_{HU}), no micro-keying of a dolomite aggregate is observed.

Table 1 Mechanical properties of two different aggregate types

AGGREGATE TYPE	HU_{plast} [GPa]	HU [N/mm^2]	E_{HU} [N/mm^2]
Polycrystalline-quartz	15604	5369	115
Dolomite	1798	1415	64

The deformation of the concrete microstructure surrounding the expansion sleeve using image analysis is presented in a composite photograph (Figure 8).

It shows the interface between the expansion sleeve (right side, grey beam) and the concrete components consisting of quartz (white) and cement matrix plus porosity (shades of grey). Each image represents an area of 1.2 x 0.8mm^2 and shows the the lower part of the sleeve-concrete interface where the main deformation took place. Areas of high deformation show significantly lower porosity than less deformed areas. Furthermore, we can observe an inner zone of high deformation with a lenticular shape approximately 0.5 mm in diameter. This inner zone is adjacent to the sleeve and has a bright grey colour. A second intermediate semicircular zone is connected to the inner area. It is characterized by the grey colour and represents a lower state of deformation, but the deformation of the concrete matrix is still detectable. Further from these two regions we observed the original undeformed concrete. The furthest deformation we could determine by this method was at a distance of approximately 2 mm from the expansion sleeve surface. The calculated quantitative results from the pore size distribution analysis of the vaying states of concrete deformation related to the area (%) of cement matrix are given in Table 2.

Table 2 Quantitive results from the pore size distribution analysis from Figure 8. The numbers represent the percentage of area within the porosity classification

AREA WITH	FIGURE 6A [%]	FIGURE 6B [%]	FIGURE 6C [%]	FIGURE 6D[%]	FIGURE 6E [%]	FIGURE 6F [%]
Low porosity	2	10	4	16	30	12
Medium porosity	22	34	22	30	32	35
High porosity	76	56	73	53	38	53

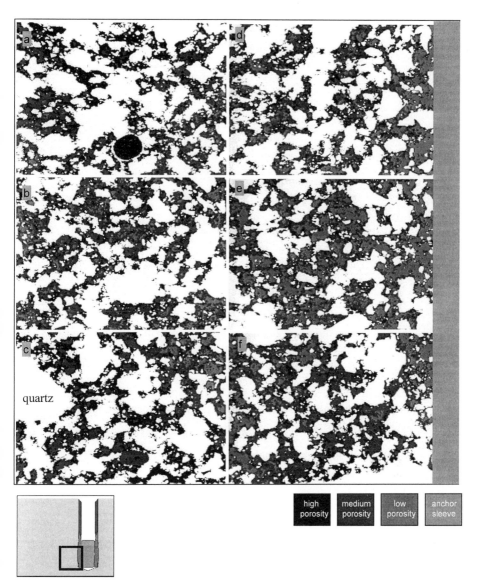

Figure 8   Anchor induced deformation of the concrete micro-structure derived by optical pore
size distribution analysis (location with respect to the anchor shown in the
small Figure at the left) [17]

The calculated radial deformation state occuring in the base material at the end of the setting
procedure equivalent to the experiment is shown in Figure 9. It should be mentioned that no
external load has yet been applied. Here regions with a high deformation state are depicted
using a dark grey colour. The main loading effects are caused by compression at the bottom
of the borehole.

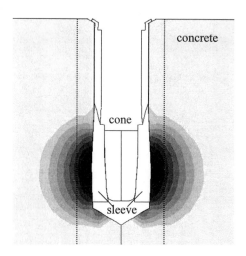

Figure 9   FEM-simulation of the deformation state of the concrete at the end of the setting procedure when no external loads has yet been applied

We checked the results from the image analysis with the solution of the FEM-simulation and found a good correlation of the distribution of deformation within the concrete microstructure. The bold black dotted lines in Figure 9 represent the distance from the expansion sleeve where concrete deformation is just measureable by image analysis. Due to the limited resolution of the image analysis system the minimum deformation of the concrete which could be measured is approximately 10 μm. Corresponding to this the FEM-simulation results predict a deformation state of 0.5-1.0 % at a distance of 2 mm from the expansion sleeve surface. This is equivalent to a deformation of 15 μm and therefore correlates well with the image analysis results.

SUMMARY

The anchoring mechanism of a HKD flush anchor in concrete was examined. Optical interface analysis and quantitative image analysis of the pore size distribution in the concrete surrounding the anchor were performed to measure the deformation state of the concrete caused by the expansion anchor.

The results of the optical analysis corresponded to an FEM-simulation of the strain distribution within the concrete.

Our results show that micro-keying between the expansion sleeves and aggregates within the concrete is an additional anchoring mechanism for these types of anchors. Besides the cement matrix properties, a further important parameter for this anchoring principle is the rigidity of the aggregate. This can be determined by universal or Vickers hardness and the derived modulus of elasticity from microscopic indentation measurements.

ACKNOWLEDGEMENTS

The authors wish to express their sincere thanks to Dr. M. Gödickemeier and Dr. S. Moseley for intensive and helpful discussions and Prof. Dr. W.J. Huppmann and Dr. E. Wisser for their continous support and interest in the work.

REFERENCES

1. SENKIW, G A, AND LANCELOT III, H B. Anchors in concrete – design and behaviour, American Concrete Institute, ACI Special Publication SP-130, Dtroit MI, USA, 1991.

2. SIA Commission 179: Fastenings in concrete and masonry, Issue 6/1998, Swiss Code No. 505179, 1998.

3. REHM, G, ELIGEHAUSEN, R. AND MALLÉE,R. Befestigungstechnik, Sonderdruck aus dem Beton-Kalender, 1992.

4. LEE, S-B. Untersuchung zur Einschlagenergie von wegkontrolliert spreizenden Dübeln, Report No. 1/66-94/20, Diplomarbeit, Institut für Werkstoffe im Bauwesen an der Universität Stuttgart, 1994.

5. LEE,S.-B. AND ELIGEHAUSEN, R. Experimentelle Untersuchungen zur Ermittlung der charakteristischen Last von unvollständig verspreizten Einschlagdübeln M8, M10 und M12 im gerissenen Beton, Bericht No 1/87-96/24, Institut für Werkstoffe im Bauwesen an der Universität Stuttgart, 1996

6. LEHMANN, R. Tragverhalten von Metallspreizdübeln im ungerissenen und gerissenen Beton bei der Versagensart Herausziehen, Dissertation, Institut für Werkstoffe im Bauwesen an der Universität Stuttgart, 1994.

7. NOGHABAI, K. Splitting of concrete covers - A fracture mechanics approach, in Fracture mechanics of concrete structures (ed F.H. Wittmann), Aedificatio Publishers, Freiburg/Breisgau, Germany, 1575-1584, 1995.

8. JUSSEL, P., WALL, F.J. AND BOURGUND, U. Application of axisymmetric finite element analysis for anchors in concrete, in Computational modelling of structures (eds H. Mang, N. Bicanic and R. de Borst), Pineridge Press, Swansea, U.K., 1047-1056., 1994.

9. NIENSTEDT, J. AND DIETRICH, C.Application of the finite element method to anchoring technology in concrete, in Fracture mechanics of concrete structures (ed F.H. Wittmann), Aedificatio Publishers, Freiburg/Breisgau, Germany, 1909-1914, 1995.

10. WITTMANN, F.H. AND HU, X. Fracture process zone in cementitious materials, International Journal of Fracture, Vol. 51, 3-18, 1991.

11.   HILLERBORG, A., MODEER, M. AND PETERSON, P.E. Analysis of crack formation and crack growth in concrete by means of fracture mechanics and finite elements, Cement and Concrete Research, Vol. 6, 773-782, 1976.

12.   VAN MIER, J.G.M. Failure of concrete under uniaxial compression: an overview, in Fracture mechanics of concrete structures (ed H. Mihashi and K. Rokugo), Aedificatio Publishers, Freiburg/Breisgau, Germany, 1169-1182, 1998.

13.   OTSUKA, K., DATE, K. AND KURITA, T. Fracture process zone in concrete tension specimens by X-ray and AE-techniques, in Fracture mechanics of concrete structures (ed H. Mihashi and K. Rokugo), Aedificatio Publishers, Freiburg/Breisgau, Germany, 3-16, 1998.

14.   NEMATI, K.M. AND STROEVEN, P. Fracture analysis of concrete: a stereological approach, in Fracture mechanics of concrete structures (ed H. Mihashi and K. Rokugo), Aedificatio Publishers, Freiburg/Breisgau, Germany, 47-56, 1998

15.   CHUY, S., CANNARD, G. AND ACKER, P. Experimental investigations into the damage of cement concrete with natural aggregate, in Brittle matrix composite (ed. A.M. Brandt and I.H. Marshall), Elsevier Applied Science, London, 341-354, 1986.

16.   HILTI AG. Fastening technology manual, B3.1-3.4 Product information, Issue 5/93, Hilti Corporation, Schaan, Principality of Liechtenstein, 1991.

17.   SCHIESSL, P AND MENG, B Untersuchung der durch einen Spreizdübel induzierten Gefügeschädigung, IBAC-Forschungsbericht F633/1, 1997.

THEME TWO:

PRACTICAL APPLICATIONS

CONCRETE STRUCTURES IN MOVABLE RIVER BEDS

J S Rocha

National Laboratory of Civil Engineering

Portugal

ABSTRACT. The presence of fixed structures imposes strong effects on riverbeds and on banks. The performance of concrete structures shall not be only analysed in the structural aspects, but also in the consequences on flow, and on modifications of riverbeds. It is a matter of fact the modern large bridges, supported by deep pile foundations, are not so vulnerable to the scours in piers; but this problem maintains its importance. The better way to tackle this problem is to consider the existence of regulations imposing some hydraulic criteria for the design of safe structures. These regulations shall consider the definition of the flow actions, the accepted risk for the water level modifications and riverbed scours, the shapes of the piers and abutments, or other bank obstacles.

Keywords: Alluvial, Riverbed, Scour, Bank, Bridge pier, Bridge abutment, Foundation, Design, Regulation.

Dr João Soromenho Rocha is a Principal Research Officer, in Hydraulics Department, National Laboratory of Civil Engineering (Laboratório Nacional de Engenharia Civil - LNEC), Lisbon, Portugal. He specialises in the alluvial river hydraulics, sedimentology, floods and water resources management. He has experience in hydraulic analysis applied on bridge design and on riverbed evolution on a large time scale.

INTRODUCTION

The riverbeds on alluvial streams have a large mobility depending on flow regime. During high flow discharges there may be large natural movements, implying vertical movements of order of a few meters in large rivers. The bed features moving horizontally during the floods superimpose the previous vertical movements with the general movements of propagation of dunes and other bed forms. The time scale of these movements may be short, say some hours, during floods, but may be also very large, during years, considering the global geologic behaviour of the river catchment.

Consequently, any fixed structure built in the riverbeds, or in the banks, is subjected to strong actions. Also, the presence of the fixed structures imposes strong effects on the riverbeds or on banks. It is common to increase the natural vertical movements of riverbeds from a fraction of a meter, to scour until a few meters. The modifications have large drawbacks, on the structure's themselves, and on the behaviour of river flow around the structure. The modifications are felt downstream as far as twenty times the width of the riverbed.

The performance of concrete structures may not be only analysed in the structural aspects, but also in the consequences on flow, and on modifications of riverbeds. The extension of performance of concrete structures built in the riverbeds, or on the riverbanks, is in this case straightforward, implying multidisciplinary teams, including hydraulic engineers.

It is a fact the modern large bridges, supported by deep pile foundations, are less vulnerable to the scours in piers; but for smaller ones, and on old bridges, this problem maintains its importance. The better way to tackle this problem is to consider the existence of regulations imposing some hydraulic criteria for the design of safe structures. These regulations consider the definition of the flow actions, the accepted risk for the water level modifications and riverbed scours, the shapes of the piers and abutments, or other bank obstacles.

Finally the water quality of river shall not be forgotten because in some cases the presence of aggressive agents must be taken into account on concrete formulation.

RIVERBED EVOLUTION

River morphology is the result of a dynamic interaction of water and sediment motion with the alluvial topography. The hydrodynamic and sediment transport processes are basically non-linear. This explains the various scale levels in their morphological behaviour [1].

The morphology of rivers involves a wide variety of space and time scales. Individual sand grains are of millimetre-size and move at the time scale of turbulence. Bed ripples and dunes are of metre size and develop at a time scale of hours, whereas the cross section shape of a channel involves spatial scales of tens to hundreds of meters and may change at a time scale of days (flood) to years (low water). The longitudinal bed profile in a channel at a scale of many kilometres may take years to centuries to adjust to a disturbance, and the channel alignment changes at a time scale of decades, centuries or even more. The pattern of tributaries at a river basin scales varies at geological time scales.

The relations between the natural river evolution and all human structures built in the river environments have implied an effort to study the natural evolution, and the consequences of the introduction of any kind of structures, obstacles to the water and sediment flow. The river hydraulics tried since along time ago to solve the problem.

The time life of each structure built in the river environment is an important criterion to choose the time scales to be analysed more carefully. The very short or very long time scales have no interest to the analysis of the behaviour of a structure in a river environment. So the time scale hours is appropriate for the crisis events like floods, when the maximum of water discharges, sediment loads and vertical movements of the riverbeds occurs. The time scales years or decades are appropriate for the analysis of a structure time life. On contrary, the time scales lower than hours or larger than centuries are not pertinent for that analysis.

Considering the useful time scales the analysis of the river evolution shall be focused in the evolution of bedforms, cross-section profiles (in straight channels and bends) and river alignments (meandering).

The term bedforms is used for smaller scale rhythmic features, such as ripples and dunes. Considering the space scale only the dunes have a practical importance for the structure analysis. The wave length, wave height, the growth rate and the migration speed are the main parameters to be quantified in each case study. The wave height controlling the vertical movements of riverbeds is maybe the most important parameter to quantify. Depending on the size of the river the dune wave height may range between a few centimetres to a few metres. But the wave height in the same cross section of the river may also range between a few centimetres to a few metres, depending on the river discharge. For instance, in Portugal the river discharge may vary more than 100 times between the long summer dry period (three months or more) and the short flood discharges (hours or days). In the average flood discharges, the Douro River, in Portugal, may migrate dunes with 3 m high.

The bank movements are maybe more important for practice of engineering than the riverbed movements. Indeed, they are a permanent dilemma with riverbanks use and natural movements of banks. Theoretically the evaluation of riverbank stability is more difficult. The movements depend on sediment transport phenomena by water (as riverbed) but also on geotechnical properties of sediment (in general in a strata status) and also on vegetation cover. Presently is increasing the research effort on a worldwide base this last factor.

The description of the complexity of the riverbed evolution make clear the difficulty to understand, the dominant factors to be considered in the civil engineering practice in the river environment.

BEHAVIOUR OF FLOW AROUND OBSTACLES

A stream with sufficient velocity scours its bed, as occurs during flood events, mainly during the rise period. The scouring process of a stream develops until the force of the water gradually decreases or the resistance of the bed increases when some sort of equilibrium is reached. This general pattern is intensified when an obstacle is inserted in a flow. The first phenomena is general scour, or degradation, the second is called local scour, or simply scour.

Local scour around piers and other obstacles in the riverbanks is the result of vortex systems that develop as the river flow is deflected around the obstacle. The main vortex system originates at the upstream nose of the obstacle where the flow acquires a downward or diving component in elevation, which reverses direction in plan at the streambed. As bed material is removed by the flow, a spiral roller develops within the hole formed, which spirals around the side of the obstacle. In plan, the vortex system has a horseshoe shape and is frequently referred to as a horseshoe vortex, Figure 1.

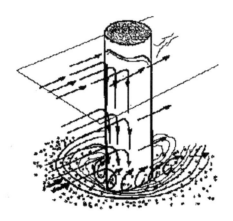

Figure 1 Horseshoe vortex formation at a cylinder pier

The scour hole will increase in size until an equilibrium depth is reached. There are two conditions: clear-water and live-bed. Clear-water means the river flow has no capacity to transport the riverbed sediment, being the average velocity lower than the critical velocity for initiating sediment movement. Consequently, the bed movement occurs only adjacent to the obstacle, where the flow velocity is increased in the vortex pattern.

In the live-bed conditions the whole riverbed is in motion, and the hole is formed by the balanced amount of sediment entering the scour hole and leaving it.

The depth at which the equilibrium condition is reached will be greatest at the transition between the clear-water and live-bed conditions, i.e. at the threshold of movement, when the approach velocity equals the average critical velocity for initiating sediment movement.

SCOUR FORECAST AND BACKWATER ANALYSIS

The complexity of the hydrodynamics associated with flow around piers, or other fixed obstacles built in riverbeds or on riverbanks, coupled with the inability of researchers to develop a generally applicable sediment transport relation have caused researches to rely on empirical relations for the prediction of scour at bridge piers and other obstacles [2].

The effect of bed material on the depth of scour is difficult to assess from most laboratory investigations because only uniform bed materials were used. Laboratory research concluded that as the gradation coefficient of the bed material increases the depth of scour decreases. However, there was little validation with field data.

Conditions in the field can be very different from conditions in the laboratory. Laboratory conditions are uniform and usually only one variable is allowed to change at a time. Conversely, conditions in the field are nonuniform and dynamic.

The research has concentrated on scour in cohesionless material and there is no guidance in the literature for estimating local scour depth in cohesive materials.

The general equation for the estimation of local scour at cylindrical pier has the form:

$$d_s = C \, U_0^m \, y_0^n \, b^q$$

where, d_s = depth of scour measured below upstream bed level
           C = experimental constant
           U_0 = approach flow velocity
           y_0 = depth upstream the pier
           b = width of the pier
           m, n, q = experimental constants.

Estimation of local scour at non-cylindrical piers may be obtained by applying suitable factors to the equations for predicting scour around cylindrical piers, using an equation like

$$d_{sg} = f_2 \, f_3 \, d_s$$

where, d_{sg} = depth of scour measured below upstream bed level
           d_s = depth of scour in a cylindrical pier
           f_2 = factor to account for pier shape
           f_3 = factor to account for oblique flow.

The factor f_3 may vary from 1 to 7, being higher values for 90° angle of attack and larger length over width ratios.

Some pier shape factors are presented in Table 1, [3].

If the obstacles have a large area the river may be confined. The effect of decreasing the free cross section of the river will be to increase the depth and velocity of flow and to make the backwater effect more severe. If the backwater effect proves excessive, adjustment of the obstacle shall be considered

The estimation of backwater is a hydraulic problem outside the presentation of this paper.

Table 1   Pier shape factor

SHAPE IN PLAN	LENGTH/WIDTH	f_2
Circular	1	1.00
Lenticular	2	0.91
	4	0.67
	7	0.41
Parabolic nose		0.80
Triangular 60°		0.75
90°		1.25
Elliptic	2	0.91
	3	0.83
Ogival	4	0.86
Rectangular	2	1.11
	4	1.40

ACTIONS ON PIERS AND ABUTMENTS

A flow against an obstacle will exert a hydrodynamic force on it, the component of which in the direction of flow is a drag action and the component normal to flow is a lift action. In fully turbulent flow, the force is generated from a combination of the shear stress against the obstacle face and the pressure differential caused by flow separation at the obstacle tail.

For both hydrodynamic forces may be used a general equation

$$F = C \, \rho \, U_0^2 \, y_0 \, L/2$$

in which
	F	= drag force F_D or lift force F_L
	C	= drag coefficient C_D or lift coefficient C_L
	ρ	= water density
	U_0	= approach flow velocity
	y_0	= depth upstream the obstacle
	L	= length representative of obstacle.

The impact forces, such as debris loading or ship collision, are other important actions to be considered in the analysis of the behaviour of obstacles in the river flows. Ship to structure collision may be a real problem where large ship traffic is present near the obstacles, bridge piers or abutments or other riverbank structure.

Human error, river currents and wind forces, poor visibility, small width of navigation spans or mechanical failures are all factors to provoke a ship or other floating material impact.

HYDRAULICS AND BRIDGE DESIGN

A bridge is one of the most important structures built in movable riverbeds. Since very old days the bridge construction was associated with a high-risk human activity. Indeed, in certain situations a bridge was considered a temporary structure.

The relation between river channel width and bridge span is always an important ratio to measure the impact on both, river and bridge. Nowadays, being the capacity to build large spans increased tremendously, compared with past centuries, the number of cases where this impact exists are minder, but not disappeared completely. Additionally, the relation span and cost is always a factor to be considered.

The concrete bridges are dominant. Even in the case of other materials the foundations are based in concrete solutions. Consequently the bridge design and concrete have strong relations and also the hydraulics is an essential subject for the bridge design.

In a typical bridge design procedure some structural, geotechnical and hydraulic factors shall be included and adjusted iteratively to achieve a bridge configuration that is satisfactory functionally, economically and aesthetically.

In past, it was obvious the selection of a bridge location considering the constraints imposed by the regime of the river. The capacity to increase the span of the bridges had attenuated the constraints, but in major part of bridges there are yet river constraints.

Among other hydraulic factors the design river flood discharge or the design flood level are maybe the most important one, and always present. Both values have a stochastic meaning. Some times it is referred the maximum flood level; that may be interpreted as the flood level associated with a determined return period. The choice of this return period shall be associated with the risk, a function of the hazard and of the vulnerability of the traffic on the bridge. For very important bridges the chosen return period may be so high as 500 years, for other minor bridges it is admissible to be flooded every year. The economic analysis shall be directly or indirectly considered in that choice.

Other parameters associated with design flood discharge are the median size of bed material, the approach flow velocity and direction, the flood plain width, the river meander characteristics.

For each bridge solution the constricted waterway width, the general and local scour depth, the backwater, the flow velocity, the training works shall be computed.

Many of the factors which will be taken into consideration in the design of a bridge crossing over a river will be common to other types of bridge. But there are three factors particularly applied to crossing over rivers: height of bridge (function of a freeboard to ensure free passage of debris or dictated by navigational requirements); the pier geometry (to consider the hydrodynamics of the flow and local scour) and pier location (to guaranty a safe backwater effect or navigation).

All these factors are important to design adequately a concrete bridge structure, being of most value a fine analysis of the bridge piers and the abutments, and all parts of the bridge that may be in contact with water. The problem is generally associated with the fact the most dangerous or critical situations occur during floods. The floods in rivers are a flash phenomena dificultting a good characterisation of all pertinent data for a safe project.

The details of the concrete structures in contact with the water, with sediments and floating materials transported by water, shall be defined considering adequate properties of the concrete to work adequately in all conditions.

RIVERBANK STRUCTURES DESIGN

The riverbank structures may have two functions. One of them is the bank protection, associated or not to river training works. Some years ago the concrete solutions were intensely used in riverbank works. Presently, according to a new look on river works, other softer solutions, like concrete in small pieces or when possible only vegetable covers, are mostly used. To adapt better to movements of the riverbanks its protections are designed with flexible solutions thus the reason to use concrete in small pieces and not in a continuous way.

The other function of river bank structures is the use of riverbank for human activities, like a mooring quay for ships, a building over the bank, an industrial facility, an intake, an outlet, an artificial canal to link waterways etc. It is practically impossible to define a small number of representative cases.

In both cases the way to consider hydraulic factors in the concrete structures design is similar to bridge case presented above. The height of the structured (compared with flood levels or with low water levels) the geometry of the structure and the relative location (in comparison with current direction) maintain the dominant importance as was the bridge case.

Maybe the additional aspect to be included is the geotechnical behaviour of the bank itself. Indeed the interface ground and waterway is relatively weak compared with other cases.

REGULATIONS

The regulations, or codes, are normal ways to guaranty the quality of a civil engineering design. Each country has being adopting the national regulations, adapted to the scientific, technical, administrative and cultural situation of that country. It is not mentioned any comparative study of the national regulations, but it is a general thought the inclusion of hydraulic factors on bridge or riverbank structures are not common. Some countries have some regulations about the choice of return period of flood discharges.

Since 1990 the CEN, European Committee for Standardization, continued the work initiated by the Commission of the Europe Communities (CEC) of establishing a set of harmonized technical rules for the design of building and civil works which would initially serve as an alternative to the different rules in the various Member States and would ultimately replace them. These technical rules became known as the "Structural Eurocodes".

The Eurocode programme is in hand on the 9 Structural Eurocodes, each generally consisting of a number of parts. The 9 Eurocodes deal respectively with:

1. Basis of design and action on structures
2. Design of concrete structures
3. Design of steel structures
4. Design of composite steel and concrete structures
5. Design of timber structures
6. Design of masonry structures
7. Geotechnical design
8. Design of structures for earthquake resistance
9. Design of aluminium alloy structures

It is not evident at a first glance the inclusion or not of the problems of hydraulic factors in these Eurocodes. If we go on the details, for instance, in Eurocode 2: Design of concrete structures - Part 2: Concrete bridges, we verify the absence of the hydraulic environment in it. Indeed, the constituent sections of that document refer to basis of design, material properties, section and member design, detailing provisions, construction and workmanship and quality control and do not mention anything about river or water flow [4].

The Eurocode 1: Basis of design and actions on structures - Part 2-4: Actions on structures - wind actions, is the only Eurocode part that present one action similar to the water action. It has 10 sections, respectively:

1. General
2. Classification of actions
3. Design situations
4. Representation of action
5. Wind pressure on surface
6. Wind forces
7. Reference wind
8. Wind parameters
9. Choice of procedures
10. Aerodynamic coefficients

There are also 3 annexes, respectively:

A. Meteorological information and national wind maps
B. Detailed procedure for in-line response
C. Rules for vortex excitation and other aerolastic effects dynamic characteristics [5]

It is not known the reason there is not prepared an equivalent Eurocode part for water actions. There is not known any statistic to explain the reason for that. Maybe one reference may indirectly explain the reason for that situation. Indeed, in the presentation of scope it is referred to "*cable stayed and suspension bridges are not covered by this Part, and specialist advice should be sought*". By analogy we may admit that all hydrodynamic action is a specialist subject.

It is noteworthy to say "*the reference wind velocity is defined as the 10 min wind velocity at 10 m above ground of terrain category II having an annual probability of exceedence of 0.02 (commonly referred to as having a mean return period of 50 years)*". For other annual probabilities of exceedence an expression is also defined.

The wind parameters are the mean wind velocity, roughness coefficient, terrain categories (4 cases), topography coefficient and exposure coefficient. All this type of parameters has an equivalent in water actions as presented above in previous sections of this paper.

If in European countries the water actions are outside the regulations for structural design, the same situation was maybe in USA. However, recent works evidence the interest in that analysis *a posteriori*. The most common cause of bridge failures is scouring of bridge foundation during floods [6]. The National Bridge Inventory of USA shows that 86 percent of the 577,000 bridges are built over waterways, and 10,000 of these bridges have been determined to be scour critical. These figures indicate a percentage of 1.7 critical to scour. Consequently, since 1988 a plan of action was developed according a Technical Advisory of Federal Highway Administration. Five steps were defined, being interesting to mention the first two of them.

The first was the creation of interdisciplinary teams of hydraulic, geotechnical and structural engineers to evaluate the scour problems of the existing bridges and in designing new. The second saying the new bridges should be designed for scour from floods equal to or less than the 100-year flood and to be scour safe for a superflood on the order of the magnitude of a 500-year flood.

The evaluation process for the scour susceptible bridges should follow the process given in HEC-18 and HEC-20. Hydraulic Engineering Circular No.18 contains the state-of-the-art methodology for evaluating scour at highway bridges. The chapters of that HEC are:

1. Introduction
2. Basic concepts and definitions of scour
3. Designing bridges to resist scour
4. Estimating scour at bridges
5. Estimating the vulnerability of existing bridges to scour
6. Inspection of bridges scour

In 1996 the status of the bridge scour program on 484,916 bridges indicated five categories, being 27.2% included in scour susceptible (131,943 bridges), 20.5% unknown foundations and 2.1% scour critical (10,276 bridges).

CONCLUSIONS

The presence of fixed structures imposes strong effects on riverbeds and on riverbanks. The performance of concrete structures shall not only be analysed in the structural aspects, but also in the consequences on flow, and on modification of riverbeds.

The presence of interdisciplinary teams of hydraulic, geotechnical and structural engineers to evaluate all problems of the existing bridges and in designing new shall be considered. It is a matter of fact the modern large bridges, supported by deep pile foundations, are not so vulnerable to the scours in piers; but this problem maintains its importance. The statistics in bridge problems in some countries show the importance of that. But the problems caused by bad bridge design in river environment shall be greater in number and not so simply assessed.

The better way to tackle this type of problem is to consider the existence of regulations imposing some hydraulic criteria for the design of safe structures. These regulations shall consider the definition of the flow actions, the accepted risk for the water level modifications and riverbed scours, the shapes of the piers and abutments, or other bank obstacles.

The water quality of river shall not be also forgotten because in some cases the presence of aggressive agents must be taken into account on concrete formulation.

The general final conclusion is that for some situations it is not enough to see only the concrete isolated but also its interaction with the environment, in both ways, concrete affected by environment, and environment affected by concrete presence. In the case of this paper, the river environment.

In the river environment it is essential to consider the form of the structures, in order to minimise the impact on the water flow and in the riverbed modification. Since, at least, Leonardo da Vinci, this fact is well known and studied. The consequence of the inclusion of an adequate form in concrete structures inserted in river environment is the need of a civil engineer action of *creating with concrete efficient and beautiful forms*.

REFERENCES

1. DE VRIEND, H J. River morphology: A manifestation of nonlinear dynamics. Managing Water. Coping with Scarcity and Abundance, Proceedings of theme A, Water for a Changing Global Community, XXVII IAHR Congress and ASCE, San Francisco, August 1997, pp.10-15.

2. MUELLER, D S, and JONES, J S. Evaluation of field and laboratory research on scour at bridge piers in the United States. Managing Water. Coping with Scarcity and Abundance, Proceedings of theme A, Water for a Changing Global Community, XXVII IAHR Congress and ASCE, San Francisco, August 1997, pp.135-140.

3. FARRADAY, R V, and CHARLTON, F G. Hydraulic Factors in Bridge Design, Hydraulics Research, Wallingford, 1983, pp 102.

4. EUROPEAN COMMUNITY FOR STANDARDIZATION. Eurocode 2: Design of concrete structures - Part 2: Concrete bridges, European Prestandard ENV 1992-2, CEN, Brussels, 1996.

5. EUROPEAN COMMUNITY FOR STANDARDIZATION. Eurocode 1: Basis of design and actions on structures - Part 2-4: Actions on structures - Wind actions, European Prestandard ENV 1991-2-4, CEN, Brussels, 1995.

6. MORRIS, J L, and PAGAN-ORTIZ, J E. Bridge scour evaluation program in the United States. Managing Water. Coping with Scarcity and Abundance, Proceedings of theme A, Water for a Changing Global Community, XXVII IAHR Congress and ASCE, San Francisco, August 1997, pp.110-116.

SERVICEABILITY BEHAVIOUR OF FRP REINFORCED CONCRETE STRUCTURES

M A Aiello

L Ombres

University of Lecce

Italy

ABSTRACT. The present paper is devoted to the analysis of the behaviour under service loads of Fiber Reinforced Plastics (FRP) reinforced concrete structures. The analysis, referred to concrete members reinforced with FRP rebars, is carried out by means of a theoretical model, derived from a cracking analysis based on slip and bond stresses. The analytical formulation of the model and the numerical procedure adopted to solve the mathematical problem are described with reference to the stabilised crack formation stage. Typical quantities that characterise the serviceability of reinforced structures such as moment-curvature and load- deflections diagrams under short and long-term loads are, then, evaluated. Finally, comparisons between model predictions, Codes predictions and experimental results are presented and discussed. The theoretical model gives satisfactory predictions of experimental results; such occurrence confirms the effectiveness of the adopted approach even if some simplifications are needed for immediate applications in the design practice. In addition, a wider experimental investigation is suggested both for deformability analysis and bond-slip laws determination.

Keywords: Fiber Reinforced Plastics (FRP), Concrete, Structures, Serviceability, Bond, Slip, Deflections

Dr M A Aiello is a Researcher in Structural Engineering, Department of Materials Science, University of Lecce, Italy. His main research interests include the structural analysis of the FRP reinforced concrete members, the adhesion systems between polymers and concrete and the analysis of composite laminated structures.

Dr L Ombres is a Research/Teaching Fellow in Structural Engineering, Department of Materials Science, University of Lecce, Italy. His main research interests include the analysis of structural behaviour of concrete elements (HSC, FRC, FRP reinforced concrete) and the analysis of sandwich laminated composite structures.

INTRODUCTION

In the field of reinforced concrete structures, use of innovative materials, as High Strength Concrete (HSC), Fiber Reinforced Concrete (FRC) and Fiber Reinforced Plastic (FRP) reinforcements, is very widespread. These structural solutions, as known, guarantee very high performances of the reinforced concrete members both from the mechanical and the durability point of view. In particular, the use of FRP reinforcement as rebars, seems to be a very effective solution to the corrosion of traditional steel rebars. The use of FRP composite materials as concrete reinforcement gives structural members with rigidity lower than that of the traditional steel reinforced concrete structures. FRP reinforced concrete structures are, then, very deformable [1].

The behaviour of FRP reinforced concrete structures is, therefore, conditioned from their high deformability; as a consequence serviceability conditions become fundamental in the structural design. For this reason, an accurate analysis that takes into account the real behaviour of materials is needed in order to evidence the advantages and benefits related to the use of innovative materials.

Starting from these considerations, the paper is devoted to the analysis of serviceability of FRP reinforced concrete flexural members as regard to deflections and cracking under short and sustained loads. A numerical procedure able to predict deflections and cracking of FRP beams is described and used in the analysis. This procedure, applicable to all FRP materials, refers to the curvature values at several sections along the member and allows deflections to be evaluated integrating the moment curvature diagrams. Non linear expressions, both for the constitutive law of compressed concrete and the bond-slip law between the concrete and the FRP reinforcement, are used and the classical hypothesis of perfect bond between concrete and reinforcement is removed. Referring to a beam element between two cracks, equilibrium on the cross-sections, compatibility of strains and bond-slip relationships, furnish a non linear system of equations that, solved by a numerical procedure, allows the curvature value to be obtained.

In the analysis, typical quantities that characterise the serviceability of r. c. beams (moment-curvature and load-deflections diagrams) are determined varying mechanical and geometrical parameters (materials strength, reinforcement ratios, surface treatment of FRP rods, etc.).

Alternative procedures, proposed from Codes for the serviceability analysis of FRP reinforced concrete beams are also used to predict flexural beam deflections. A comparison between results furnished by the adopted procedure, Codes relationships and experimental tests on Glass Fiber Reinforced Plastic (GFRP) and Aramid Fiber Reinforced Plastic (AFRP) beams, allows us to evidence that in many cases a general method, such as that adopted, is needed for an accurate serviceability analysis. Additionally the comparison shows good agreement between results predicted by the adopted procedure and those experimental.

MODEL FOR SERVICEABILITY ANALYSIS

The general model adopted for the serviceability analysis of flexural concrete members reinforced with Fiber Reinforced Plastics rods, is derived from a cracking analysis of structural members founded on slip and bond stresses [2], [3]. The analysis refers to the stabilised crack formation phase; in this situation, as known, all crack spacing vary between l_b and $2l_b$, l_b being the development length.

The two cracking configurations, corresponding to the maximum and the minimum spacing, $l_{max}=2l_b$ and $l_{min}=l_b$ respectively, are considered in the model; all possible cracking configurations are between those analysed. In the first case (MCS), $l_{max}=2l_b$, the slip is zero at the half-way between cracks and the tensile concrete stress σ_{ct} reaches the tensile strength f_{ct} at the extreme fibre of the same cross-section (Figure 1).

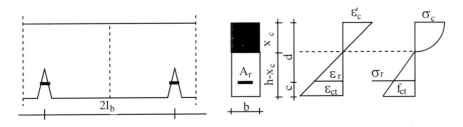

Figure 1   Maximum crack spacing configuration: stresses and strains at the half-way ($s=0$)

When the crack spacing is minimum (mcs), $l_{min}=l_b$, because of the symmetry condition the slip is zero at the halfway between cracks while the tensile concrete stress is unknown but less than f_{ct} at the extreme fibre of the cross-section (Figure 2).

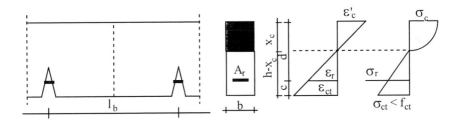

Figure 2   Minimum crack spacing configuration: stresses and strains at the half-way ($s=0$)

Analytical relationships of the adopted procedure are defined with reference to a beam element between cracks subjected to a constant bending moment higher than the first cracking bending moment. Considering a segment length dz, analytical relationships are defined by following equilibrium and compatibility conditions (Figure 3):

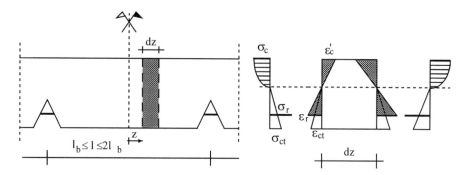

Figure 3 Block between cracks: stress and strain distribution at the ends of the segment dz

1. Equilibrium for FRP rebars

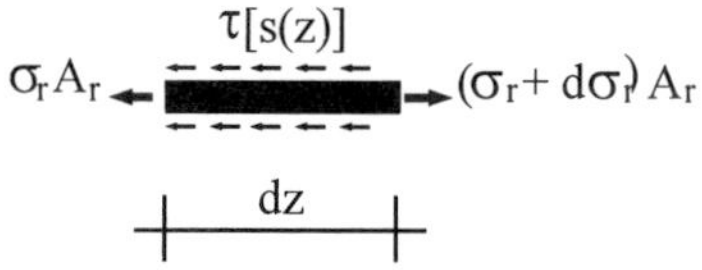

Figure 4 Stresses of FRP rebars

$$\frac{d\sigma_r}{dz} = \frac{A_r}{p_r}\tau[s(z)] \tag{1}$$

p_r and A_r being the perimeter and area of the reinforcement, respectively

2. Compatibility condition between two points, initially superimposed, belonging to the FRP rebars and the concrete

$$\frac{ds}{dz} = \varepsilon_r(z) - \varepsilon'_{ct}(z) \tag{2}$$

In the Eqns (1) and (2) σ_r and ε_r are the stress and strain of FRP rebars, ε'_{ct} is the strain value of the concrete in tension at the reinforcement level and $s(z)$ is the slip between the FRP rebars and the concrete.

3. Equilibrium conditions of cross-sections at the ends of the segment

$$\int_{\Omega_c}\sigma_c\,d\Omega_c + \sum\sigma_{ri}A_{ri} = 0$$

$$\int_{\Omega_c}\sigma_c\,y_c\,d\Omega_c + \sum\sigma_{ri}\,y_{ri}\,A_{ri} = M \tag{3}$$

Ω_c being the concrete cross-section, A_{ri} the area of the FRP rebar, y_c and y_{ri} the distance from concrete fibers and FRP rebars to the centroidal axis of the cross-section, respectively.

Equations (1), (2) and (3) furnish a system of differential equations, which can be solved by means of numerical procedures. In the paper a numerical procedure founded on the finite differences method is used and the block between cracks is subdivided in discrete elements length Δz_i. The numerical procedure starts evaluating stresses distribution at the halfway cross-section by equilibrium Equation (3) and imposing the condition $s=0$ (i.e. supposing the cross-section planarity, see Figures 1 and 2).

Later on, Equations (1), (2) and (3) evaluate the stress and strain distributions at the end cross-section of the first discrete element contiguous to the halfway cross-section. At this stage a linear distribution of strains is supposed but compatibility is given only between reinforcement in tension and the concrete in compression (Figure 3).

Subsequently the procedure is extended to every discrete element and it halts when the concrete tensile strain becomes zero; this situation, in fact, corresponds to a crack opening. The problem solution allows defining both stress and strain for the FRP reinforcement and the concrete and the bond stress-slip distribution between cracks. Additionally the development length l_b is evaluated as $l_b = \Sigma_{i=1,n} \Delta z_i$ being n the number of discrete elements between the halfway and the cracked cross-section. The curvature value of the block is obtained as the sum of the curvature values of every discrete element

$$\chi = \frac{1}{l_b} \sum_{i=1}^{n} \left(\frac{\varepsilon_r - \varepsilon_{cl}'}{d} \right)_i \Delta z_i \tag{4}$$

being d the effective depth of the beam. The χ values are, then, assumed as mean curvature values for each block; by the above-described procedure, it is possible to evaluate the moment-curvature diagram, varying the bending moment. The integration of the moment-curvature diagram along the beam axis allows evaluating beam deflections.

Bond Stress-Slip Model

As above mentioned, the adopted model used for the serviceability check of FRP reinforced concrete structures, is based on a bond stress-slip analysis. The bond behaviour, as known, is a critical aspect of the structural behaviour of FRP reinforcements both at the serviceability and at the ultimate state. In particular, at the serviceability limit state many verifications, such as the control of the cracks amplitude and structures deformations, involve evaluation of the contribution of the tension stiffening that directly arises from bond. The bond stress-slip relationships are, generally, derived from pullout tests or by beam tests; experimental results evidence a non-linear bond behaviour of FRP reinforcements.

Because of FRP reinforcements are made using different kinds of fibres (glass, carbon, kevlar, etc.) and their outer surfaces can present various shapes (braided, grain covered, intended, ribbed, etc.), the bond-slip law is strongly conditioned from each thecnological solution. As a consequence a correct bond analysis requires an experimental investigation in order to define the bond behaviour of the adopted FRP reinforcement.

Recently this problem was analysed by Cosenza et al. [4]; a detailed investigation was carried out and a general model of the bond stress-slip of FRP reinforcements was proposed. This model was obtained by a modification of the Bertero-Eligehausen-Popov law by means of a wide analysis of available experimental data; analytically it is defined by following relationships (Figure 5):

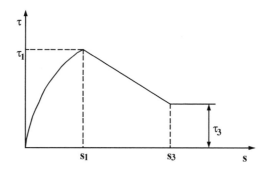

Figure 5 The bond stress-slip law for FRP reinforcements

$$\frac{\tau}{\tau_1} = \left(\frac{s}{s_1}\right)^\alpha \text{ if } s \leq s_1 ; \qquad \frac{\tau}{\tau_1} = 1 - p\left(\frac{s}{s_1} - 1\right) \text{ if } s_1 \leq s \leq s_3 ; \qquad \tau = \tau_3 \quad \text{if } s \geq s_3$$

The coefficients α, p and τ_3 must be calibrated experimentally and are depending, as Cosenza et. al. demonstrate [4], on the outer surface shape of the FRP reinforcements; some values of these coefficients, derived from a statistical analysis of experimental data, are reported in [4].

RESULTS AND DISCUSSION

Predictions of the adopted general procedure, above described, are compared with those furnished by ACI [5] and JSCE [6] Codes and with experimental results obtained both from Authors, by flexural tests on AFRP r.c. beams, and from data available in the literature [7], [8]. Beams tested from authors have been made with Portland cement concrete (cement: 400 kg/m³, sand: 810 kg/m³, aggregate: 1050 kg/m³, super plasticizier: 0.15 l/m³), and reinforced with Aramid Fiber Reinforced Plastic, called Arapree, having a diameter of 7.5 mm and sanded on the surface in order to improve the bond with the surrounding concrete.

For theoretical predictions, according to the Eurocode EC2, a non linear constitutive laws of the concrete in compression is used while constitutive laws of the concrete in tension and the FRP rebars are considered as elastic-linear [3]. Moment and load versus deflections diagrams of FRP r. c. beams are reported in the Figures 6 and 7 while in the Figure 8 load versus crack width diagrams of the midspan section of a r.c. beam are drawn. The tested beams, with rectangular cross-section, are simply supported with span L and subjected to two symmetrical forces.

Details of the data of tests are reported in Table 1 together with the reinforcement ratios $\mu=A_r/bd$ (A_r is the area of the FRP rebars, b and d are the width and the effective depth of the cross-section, respectively) and the mechanical characteristics of materials: the mean cylindrical strength of the concrete in compression f_c, the strength and the elastic modulus of the FRP in tension f_{ru} and E_r. Experimental curves, drawn in Figures 6 and 8, are reproduced graphically from data reported in [7] and [8], respectively.

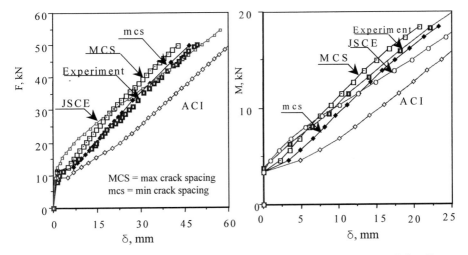

Figure 6 AFRP r.c. beam: load-deflections diagrams.

Figure 7 GFRP r.c. beam [7]: bending moment-deflections diagrams .

Comparisons show a good agreement, particularly for serviceability stresses (0.3-0.5 of the ultimate stresses), between experimental results and predictions of the adopted model that seem to be more accurate than Codes predictions. However, a more wide comparison with experimental data is needed in order to validate the effectiveness of the adopted model even if it is more onerous from a computational point of view than Codes models and, then, not immediately applicable for practical purposes.

Table 1 Data of the analysed tests

TEST	FRP	b (mm)	h(mm)	d(mm)	A_r (mm²)	μ (%)	L (mm)	f_c(MPa)	f_{ru}(MPa)	E_r(MPa)
authors	AFRP	150	200	165.00	265.07	1.07	2610	46.2	1506	50100
[7]	GFRP	130	180	147.85	237.65	1.24	1500	53.1	773	38000
[8]	GFRP	500	185	145.00	886.74	1.22	3400	30	770	42000

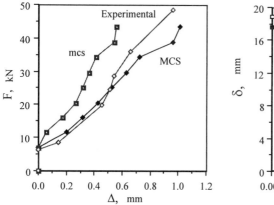 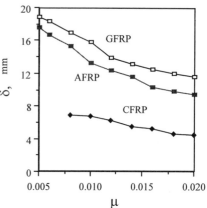

Figure 8 Cracks width of the midspan section
Comparisons between predicted and experimental [8]
results

Figure 9 δ–μ curves varying the
kind of FRP rods (b=150 mm,
h = 250mm, d= 220 mm)

Predictions of the adopted model, considering the maximum spacing cracks configuration,
are shown in the Figures 9 and 10. Curves drawn in these figures evidence the influence of
the reinforcement ratio on the deformability of FRP reinforced concrete beams; in the
Figure 9 curves are relative to different kind of FRP rebars in presence of short term loads
while in the Figure 10 curves are drawn for GFRP r.c. beams in presence of long-term loads.
Results shown in the Figures 9 and 10 refers to a simply supported beam with span $L= 2500$
mm, having rectangular cross-section and subjected to a two symmetrical and equal forces
$F=0.5\ F_u$ being F_u the load carrying capacity.

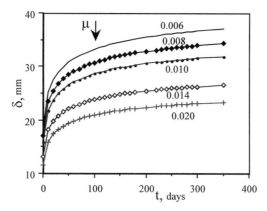

Figure 10 Long term deflections of GFRP r.c. beam with rectangular cross-section (b=150
mm, h= 250mm, L= 2500 mm; d= 220 mm; F= 0.5 F_u)

Obtained results confirm the increase of the stiffness of the FRP reinforced concrete beams with the reinforcement ratio and show the influence of the creep on the deformability of such structures.

CONCLUSIONS

1. The serviceability behaviour of FRP reinforced concrete flexural beams is well evaluated by means of a model based on a bond stress-slip analysis. The general procedure, adopted in this study, allows us to take into account all parameters that influence the serviceability behaviour of FRP r.c. structures both for short and long-term loads.

2. Comparisons between predictions of the adopted procedure and experimental results evidence a good agreement. A more extensive investigation is, however, needed to validate the procedure.

3. The serviceability analysis of FRP flexural structures has to consider the effects of the slip between the FRP reinforcements and the concrete. The analytical formulation of the bond stress-slip is essential and has to be defined accurately considering all parameters affecting the interaction between concrete and FRP rebars.

4. For practical purposes, the procedure here adopted couldn't be immediately applicable because it is onerous from a computational point of view. However, obtained results can be used to improve traditional design models.

REFERENCES

1. NANNI, A. Fiber-Reinforced Plastics (FRP) reinforcement for concrete structures; properties and applications. Elsevier Ed., 1993.

2. AVALLE, M, FERRETTI, D, IORI, I AND VALLINI, P. Sulla deformabilità di elementi inflessi in conglomerato armato. Proc. of the 10th Nat. Conf. of the CTE, Milano, 1994, pp. 33-42.

3. AIELLO, M A AND OMBRES, L. Flexural behaviour of FRP reinforced concrete beams under service conditions. Proc. Third Int. Symp. on Non-metallic (FRP) reinforcements for Concrete Structures, Sapporo, Japan, Vol. 2, 1997, pp. 405-412.

4. COSENZA, E, MANFREDI, G AND REALFONZO, R. Behaviour and modeling of bond of FRP rebars to concrete. Jour. of Comp. for Construction, 1997, Vol. 1, No. 2, May, pp. 40-51.

5. ACI COMMITTEE 440   State - of - the - Art Report on Fiber Reinforced Plastics reinforcement for concrete structures, 1996.

6. JSCE   State of the art report on continuous fiber reinforcing materials. Concrete Engineering Series 3, Tokyo, 1993.

7. THERIAULT, M, AND BENMOKRANE, B. Effects of FRP reinforcement ratio and concrete strength on flexural behaviour on concrete beams. Jour. of Comp. for Construction, 1998, Vol. 2, No. 2, February, pp. 7-16.

8. PECCE, M, MANFREDI, G AND COSENZA, E. Experimental behaviour of concrete beams reinforced with glass FRP bars. Proc. of the ECCM-8, Naples, Italy, Vol. 1, 1998, pp. 227-234.

STRENGTHENING OF CANTILEVER CONCRETE SLABS USING GFRP STRIPS

J G Teng   L Lam

Hong Kong Polytechnic University

W Chan

China State Construction Engineering Corporation

J Wang

L & M Specialist Construction Ltd

Hong Kong

ABSTRACT. Despite many studies on beams and slabs strengthened using FRP plates, no study appears to have been conducted on the strengthening of cantilever concrete slabs (canopies, balconies etc) using FRP materials. This paper presents the results of an experimental study into the feasibility and effectiveness of strengthening deficient concrete cantilever slabs by bonding glass fibre-reinforced plastic (GFRP) strips on the top surface (the tension side). As the key to the success of this strengthening method is a proper way of anchoring the GFRP strips into the supporting wall and the slab, the effectiveness of different anchorage systems was the focus of the experiments. Based on the test results, a simple and effective method is identified, in which the GFRP strips are anchored into the walls through horizontal slots and into the slab using fibre anchors.

Keywords: GFRP strips, Cantilever slabs, Strengthening, Debonding, Fibre anchors

Dr J G Teng is an Associate Professor in the Department of Civil and Structural Engineering, The Hong Kong Polytechnic University, Hong Kong, China. His current research interests include structural retrofitting using FRP composites and steel shell structures. He has authored/co-authored about 80 papers, over half of which are refereed journal papers.

Mr L Lam is a PhD student in the Department of Civil and Structural Engineering, The Hong Kong Polytechnic University, Hong Kong, China. His main research interests are in concrete technology and structural retrofitting using FRP composites.

Mr W Chan is an Assistant Engineer with China State Construction Engineering Corporation, Hong Kong, China.

Dr J Wang is the Senior Resident Manager of L & M Specialist Construction Ltd, Hong Kong, China. His main interests cover concrete durability, and repair and strengthening of concrete structures.

INTRODUCTION

Strengthening of reinforced concrete structures using fibre reinforced plastic (FRP) materials has been a topic of strong research interest in recent years. The method offers many advantages over the conventional steel plate bonding technique, including a high strength/weight ratio and high corrosion resistance. Many studies on beams, slabs and columns strengthened using FRP plates have been carried out in recent years [eg 1-6]. However, no study appears to have been conducted on the feasibility and effectiveness of strengthening cantilever concrete slabs (concrete canopies and balconies) using FRP materials. This paper presents such a study.

Cantilever concrete beams and slabs are common features in buildings. In Hong Kong, the use of concrete canopies and balconies is widespread. As concrete cantilever slabs are basically one-way slabs and hence statically determinate structures, they can fail suddenly once their moment capacity is exceeded at the support. There have been a number of failures of cantilever slab structures in Hong Kong in recent years, involving injuries to and deaths of people, causing major public concerns over the safety of such structures. Most of these failures were due to the deterioration or mispositioning of steel reinforcement. The development of a simple and effective method to rehabilitate deficient concrete cantilever slabs is thus of major significance. Although bonding of steel plates has been used widely to strengthen beams and slabs, there has been no report in the open literature on the use of steel plate bonding to cantilever concrete slabs. Since the top surface of cantilever slabs is generally exposed to weather attack, corrosion is a serious problem with steel plate bonding, particularly in a coastal environment like Hong Kong. Furthermore, there are also difficulties in site handling of steel plates as the space available for retrofitting a cantilever slab is usually limited. The use of FRP composite materials is therefore an attractive option.

This paper presents a brief summary of the results of an experimental study into the feasibility and effectiveness of strengthening deficient concrete cantilever slabs by bonding glass fibre-reinforced plastic (GFRP) strips on the top surface (the tension side). A detailed description of the study can be found elsewhere [7]. The deficiency of laboratory model slabs was induced by placing the steel reinforcement in the compression zone. This simulates the mispositioning of steel bars, a common problem for real structures. Since the key to the success of this strengthening method is a proper way of anchoring the GFRP strips into the supporting wall and the slab, the effectiveness of different anchorage systems was the focus of the experiments. Based on these results, a simple and effective strengthening method is identified, in which the GFRP strips are anchored into the walls through horizontal slots and into the slab using fibre anchors.

EXPERIMENTAL DETAILS

Experimental Set-Up

The experimental work of this study included seven tests (Tests I – VII) on cantilever slabs supported on a short wall segment. Six model slabs of identical dimensions and reinforcement details were cast, with the same slab being used in Test II and Test III. That is, the damaged slab of Test II was strengthened with GFRP strips again as Specimen III, with the strips anchored into holes in the wall. All tests were carried out by fixing the concrete specimen on a stiff support erected on the floor. The slab was positioned with its plane being vertical so that loading could be applied more easily using a hydraulic jack fixed on the floor (Figure 1). The bottom edge of the specimen was placed slightly above the floor level to avoid any contact between the test specimen and the floor. A line load was applied near the free end of the slab and was at 600 mm from the fixed end of the slab (ie the wall surface).

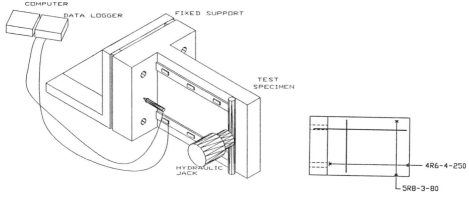

Figure 1 Experimental set-up   Figure 2 Reinforcement details

Test Specimens

The specimen is composed of two main parts: the slab and the wall. The length of the model cantilever slabs is 700 mm (from the wall surface to the free end). The thickness and width of the slabs are 100 mm and 500 mm respectively. The steel reinforcement details are shown in Figure 2. The cantilever slab and the wall were cast together to simulate real situations closely. The wall was designed to be so strong that no damage/failure of the wall would occur during testing. The wall simply offered a support to the slab and its strength was not studied. The slab was deliberately made deficient through the 'mispositioning' of the tension reinforcement as the compression reinforcement with 20 mm concrete cover, which may occur in real structures due to construction errors. As a result, the slab was unable to resist the load intended for it to carry.

Strengthening using GFRP Strips

The deficient cantilever slab, after proper curing, was then strengthened by two GFRP strips, each of 80 mm in width. The outer longitudinal edge of each strip was at a distance of 65 mm from the edge of the slab. The GFRP strips were constructed by saturating fabric strips with epoxy. The woven fabric consists of longitudinal glass fibres with transverse Kevlar fibres. The two components of epoxy resin were mixed with a mechanical mixer except for Test II for which manual mixing was used. The saturated fabric strips were laid by hand to the prepared slab surface by subjecting it to a uniform tensile force and assuring a smooth final appearance. An additional thin layer of well mixed epoxy was finally applied on the fabric. The GFRP composite strips were then left for curing for at least 7 days before testing. The cured composite strips were of uniform thickness and density with all fibres properly covered by epoxy and showed no signs of porosity. Both the fabric and the two-component epoxy were supplied by L and M Specialist Construction Ltd and form the proprietary product known as Tyfo S Fiberwrap.

Loading Procedure and Instrumentation

The load was applied at a rate of about 0.5 kN/minute and strain gauge and displacement transducer readings were taken at an interval of 0.2 to 0.4 kN in the elastic stage. However, after the first peak load (judged from the load deflection curve), the load generally decreased by a significant amount and then started to increase again but only slowly, although deflections continued to increase rapidly. Therefore, test control was based on deflections after the first peak load. Data readings were then taken at a displacement interval of about 0.3 mm based on the deflection at the loading position, but much larger intervals were used in the final stage of the tests when the deflections were very large.

Displacement transducers were installed at three locations along the mid-width line of the slab at 30 mm, 300 mm, 600 mm respectively from the wall surface. Three additional transducers were installed on the supporting wall to check the rotation and lateral displacement of the supporting wall. Only the net deflection at the loading position (ie at 600 mm from the wall surface) is presented in this paper. This net deflection excludes any deflection at the loading position due to rotations and lateral movements of the supporting wall. Stain gauges were also installed at various locations. For all the specimens strengthened with GFRP strips, three strain gauges approximately located at 30 mm (near the fixed end), 300 mm (at the mid-span) and 550 mm (near the loading position) respectively from the wall surface were installed on at least one of the two strips and in Specimens IV-VII on both strips. In addition, the strains on the compressive surface near the fixed end (at approximately 30 mm from the wall surface) were measured at the mid-width of the slab. The readings of the strain gauges and transducers were directly recorded by a PC through a data-logger.

Test Programme

Table 1 lists the seven slab specimens. Specimen I is a control specimen without strengthening using GFRP strips. Specimens II-IV were strengthened with GFRP strips with different anchorage systems to the wall, while Specimens V-VII were strengthened with GFRP strips with different anchorage details to the slab in addition to horizontal slot anchorage to the wall.

Material Properties

Three concrete cubes (150 x 150 x 150 mm) for each of the model slab test specimens were prepared for uniaxial compression tests. The average compressive strength from cube tests for each model slab is given in Table 1. Besides, concrete tensile strength and the initial Young's Modulus were also determined for the concrete of Specimen I (the control specimen) by conducting two splitting tests and two uniaxial compression tests respectively on concrete

Table 1 List of seven model slab specimens

SPECIMEN	ANCHORAGE TO THE WALL	ANCHORAGE TO SLAB	COMMENTS	COMPRESSIVE STRENGTH (MPa)
I	Not Applicable	Not Applicable	Control Specimen	38.96
II	External Wrapping	None	The same concrete	51.83
III	Inclined Slots	None	slab was used	51.83
IV	Horizontal Slots	None		49.28
V	Horizontal Slots	Wrapping around Slab Edge		51.58
VI	Horizontal Slots	Fibre Anchors		51.03
VII	Horizontal Slots	Fibre Anchors	2 GFRP Layers	46.95

cylinders (diameter = 100 mm, height = 200 mm). The resulting average tensile strength = 2.88 MPa and average initial Young's modulus = 29.3 MPa. The tensile strength of the GFRP composite averaged from five specimens is 438 MPa, assuming a nominal thickness of 1.27 mm as suggested by the supplier. The Young's modulus and Poisson's ratio averaged from two tests are 24.6 GPa and 0.275 respectively. The rupture strain can thus be calculated to be 1.78% by dividing the tensile strength by the elastic modulus. Properties of other materials involved in the tests are given elsewhere [7].

RESULTS AND DISCUSSION

Summary of Results

Seven tests including one control test were carried out in this project. For most tests, the load deflection curve exhibits an initial peak load, and an ultimate load (Figure 3).

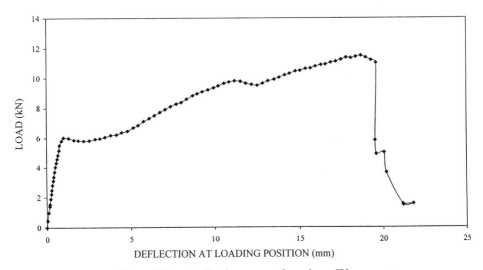

Figure 3 Load-deflection curve of specimen IV

Both loads and their corresponding deflections and strains are summarized in Table 2. Since Specimens I (the control specimen) and II failed in a brittle manner, their initial peak loads are also the ultimate loads. In addition, the initial peak is not so obvious on the load-deflection curve of Test III as the specimen was already damaged before it was strengthened by GFRP strips.

Table 2 Summary of test results

	TEST I	TEST II	TEST III	TEST IV	TEST V	TEST VI	TEST VII
Initial Peak Load P_I (kN)	3.89	6.83	7.72	6.90	7.03	6.02	9.35
Deflection at Loading Position under P_I (mm)	0.70	1.50	8.56	1.10	1.00	1.01	1.32
Average Strain in GFRP Strips near the Fixed End under P_I		0.012% (from upper strip only)	0.68%	0.033%	0.048%	0.036%	0.049%
Ultimate Load P_u (kN)	3.89	6.83	8.73	9.38	10.63	11.51	15.11
Ultimate Load / Ultimate Load of Test I	1.00	1.76	2.24	2.41	2.73	2.96	3.88
Deflection at Loading Position under P_u (mm)	0.70	1.50	32.45	21.62	52.36	18.71	16.71
Average Strain in GFRP Strips near the Fixed End under P_u		0.012% (from upper strip only)	0.87%	0.85%	1.12%	1.14%	0.90%
Rupture Strain in Upper Strip					1.14% near fixed end	1.50% near fixed end	
Rupture Strain of Lower Strip					1.44 near loading position	1.61% near fixed end	
Failure Mode	Concrete Tensile failure	Failure of Anchorage to Wall	Debonding of GFRP Strips	Debonding of GFRP Strips	Rupture of GFRP Strips	Rupture of GFRP Strips	Fibre Anchor Failure
Maximum compressive strain in concrete near fixed end					0.397%	0.176%	0.30%

Initial Peak Load and Initiation of Debonding

Except for the control specimen and Specimen II which failed in a brittle manner, the failure process in all other specimens was started by the debonding of GFRP strips from the concrete slab near the fixed end [7]. For a specimen without pre-damage, the initial peak load was found to be closely associated with the initiation of debonding. The initiation of debonding is believed to be the result of high interface stresses between the concrete and the GFRP strips caused by the formation of cracks, although before the initial peak load, these cracks could hardly be seen. Soon after the initial peak load, a major crack formed across the width of the slab and debonding was initiated. The behaviour of Specimen III is somewhat different with regard to the initiation of debonding as a result of a modified interfacial stress state due to the pre-existence of cracking. As the initiation of debonding starts only after the crack has reached a certain size, the initial peak load (debonding load) is thus significantly higher than the tensile failure load of the control specimen, and can be increased significantly by adding an additional layer of composite.

Propagation and Arrest of Debonding

Once debonding from the slab is initiated, it then propagates toward the free end of the slab. Specimens III and IV failed when complete debonding of the GFRP strips from the concrete slab was achieved. The load deflection curve of Specimen IV is shown in Figure 3, which indicates that during the propagation of debonding the deflection increased rapidly but the load carrying capacity increased only slowly. The two big drops in the load deflection curve (Fig. 3) soon after reaching the ultimate load of 9.38 kN are due to the different debonding rates of the two GFRP strips due to unintended unsymmetry in loading, specimen geometry and material distribution. The tensile strength of the GFRP strips was thus not fully utilized.

For Specimen V, the GFRP strips were wrapped around the free edge of the slab. As a result, complete debonding was not observed as debonding could not propagate around the free edge. This improved the strength significantly, and also changed the mode of failure to tensile rupture of the GFRP strips. The full utilization of the tensile strength of the GFRP strips was, however, achieved at the expense of large deflections. Furthermore, wrapping the GFRP strips around the free edge may not be economical if the span of the cantilever slab is large. Therefore, in Test VI, fibre anchors supplied by L & M Specialist Construction Ltd were installed near the fixed end of the slab to arrest propagating debonding. The fibre anchor consists of two parts (Figure 4a): the anchor bolt and the protruding fibres. In the installation, the fabric strips saturated with epoxy was first applied on the slab surface, and then the anchor bolts were inserted through the wet epoxy-fabric composite into holes of 10 mm in diameter pre-drilled on the slab. The protruding fibres were next bent and spread out on the upper surface of the composite strip (Figure 4b). In this test, an additional layer of composite was added on the composite layer formed from the protruding fibres of the anchor to enhance the anchorage capacity as is normally recommended for such fibre anchors. Two fibre anchors were installed on each strip at 150 mm and 300 mm from the wall surface respectively. Debonding was still observed to appear first between the fixed support and the first fibre anchor and then between the first and second fibre anchor (Figure 5a), with the corresponding formation of a major crack in each zone, but it did not propagate beyond the second fibre anchor (Figure 5b). The specimen eventually failed by tensile rupture of the

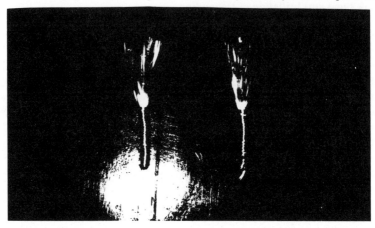

(a) Before Installation

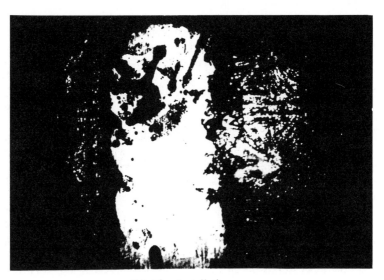

(b) After Installation

Figure 4 Fibre anchors

GFRP strips (Figure 5b). The fibre anchors thus led to an increase in the load-carrying capacity and a reduction in the deflection (Figure 6). The fibre anchors therefore provide an effective means to arrest the propagation of debonding.

The fibre anchors were unable to arrest the propagation of debonding when two layers of GFRP strips were used in Specimen VII. Part of the reason may be that the covering layer above the anchors was omitted in Specimen VII. Nevertheless, this demonstrates that the loading carrying capacity of the fiber anchors needs to be understood before they can be used.

(a) Development of the Second Crack

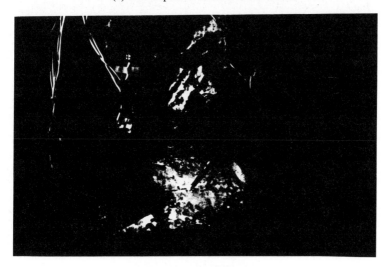

(b) Rupture of GFRP Strips

Figure 5 Failure process of specimen VI

Ultimate Load and Rupture Strain

The ratio between the failure load of the strengthened slab and the control slab is given for each case in Table 2. Significant strength enhancement is observed in all cases. For Specimens III and IV which failed by debonding of the GFRP strips from the slab, the ultimate loads are more than twice that of the control specimen. For Specimen V and VI which failed by tensile rupture of the GFRP strips, the failure loads are nearly three times that of the control specimen. With two layers of composite strips, the failure load of Specimen VII is almost four times that of the control specimen. Although a small amount of the strength enhancement was due to the slightly weak concrete used for the control specimen, most of the strength gain is due to the GFRP strips. The measured strains of the GFRP strips nearest to the location of rupture are between 1.1% and 1.6 (see Table 2), which are below the average value of 1.78% from tensile tests. Two factors may be responsible for the differences: (a) The strain gauges were not exactly at the locations of rupture failure; and (b) Some bending is present in the GFRP strips which may reduce the tensile rupture strain.

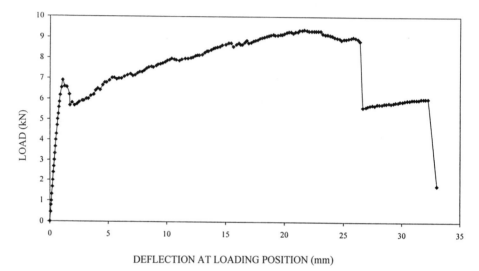

DEFLECTION AT LOADING POSITION (mm)

Figure 6 Load-deflection curve of specimen VI

Compressive Strain in Concrete near the Fixed End

Compressive strains near the fixed end (30 mm from the wall surface) were measured at the mid-width of the compressive side of the concrete slab in Tests V, VI and VII. This compressive concrete strain in Test V reached 0.397%, signifying that concrete compression failure was being approached although this was not observed in the test. This compressive strain was reached with a load of 8.30 kN (the ultimate load is 10.63 kN for this specimen), after which the strain started to decreased. It is not clear why this happened, but a possible cause may be the debonding of the strain gauge from the concrete. Comparatively small compressive concrete strains were developed in Test VI at failure. By adding another layer of composite material, larger compressive strains were developed in the concrete in Test VII.

CONCLUSIONS

Six cantilever slabs without tension reinforcement were strengthened using GFRP composite strips and then tested to failure to investigate the effectiveness of this method of strengthening. The test results have demonstrated that a large increase in the ultimate load and ductility can be achieved if the slot anchorage system is used to anchor the strips into the supporting wall. The effect of this strengthening method is even better if fibre anchors are installed to anchor the strips into the slab or the free ends of GFRP composite strips are wrapped around the free edge and onto the soffit of the slab. In conclusion, GFRP strips with slot anchorage to the supporting wall and with fibre anchors to prevent debonding failure from the slab provide a simple and effective system to strengthen deficient cantilever slab structures. A number of issues remain to be investigated for this method to be used widely. Firstly, the debonding mechanism, including its initiation and propagation, needs to be carefully studied and better understood. Secondly, the strength of the fibre anchor in resisting combined tension and shear should be studied. Thirdly, the capacity of the slot anchorage system needs to be further clarified. Finally, a simple method to estimate the strength of slabs strengthened using this method needs to be developed. Nevertheless, the method is readily applicable in practice, provided model tests are carried out for each individual case.

ACKNOWLEDGEMENT

The GFRP materials used in the tests were provided by L & M Specialist Construction Ltd. The authors are grateful for this support.

REFERENCES

1. GARDEN, H N, HOLLAWAY, L C AND THORNE, A M. A preliminary evaluation of carbon fibre reinforced polymer plates for strengthening reinforced concrete members. Structures and Buildings, ICE Proceedings, Vol 123, 1997, pp 127-142.

2. MALEK, M, SAADATMANESH, H AND EHSANI, M R. Prediction of failure load of R/C beams strengthened with FRP plate due to stress concentration at the plate end. ACI Structural Journal, Vol 95, No 2, 1998, pp 142-152.

3. MEIER, U AND WINISTORFER, A. Retrofitting of structures through external bonding of CFRP sheets. Non-Metallic (FRP) Reinforcement for Concrete Structures, Ed. L. Taerwe, E & FN Spon, 1995, pp 465-472.

4. SHARIF, A, AL-SULAIMANI, G J, BASUNBUL, I A, BALUCH, M H AND GHALEB, B N Strengthening of initially loaded reinforced concrete beams using FRP plates. ACI Structural Journal, Vol 91, No 2, 1994, 160-168.

5. SPADEA, G, BENCARDINO, F AND SWAMY, R N. Structural behaviour of composite RC beams with externally bonded CFRP. Journal of Composites for Construction, ASCE, Vol 2, No 3, 1998, pp 132-137.

6. TOUTANJI, H, AND BALAGURU, P. Durability characteristics of concrete columns wrapped with FRP tow sheets. Journal of Materials in Civil Engineering, ASCE, Vol 10, No 1, 1997, pp 52-57.

7. TENG, J G, LAM, L, CHAN, W AND WANG, J. Retrofitting of Deficient RC Cantilever Slabs using GFRP Strips, to be published.

PERFORMANCE OF CONCRETE IN CONTAMINATED LAND

S L Garvin

J P Ridal

M Halliwell

Building Research Establishment

United Kingdom

ABSTRACT. Contaminated land sites are increasingly being redeveloped for housing, industrial, commercial and leisure use. The potential risks from contamination in the ground extend from human health to damage to buildings. BRE has been undertaking a research project to assess the potential for contaminants in the ground to attack concrete. This paper reports the results of concrete cubes that have been immersed in solutions of contaminants for three years. The results suggest that some concrete mixes are likely to be degraded to a greater extent by the mixture of contaminants as opposed to sulfate acting on its own. However, mixes that are recognised as being resistant to contaminant attack have performed well to three years. In addition, beneficial effects of a water-reducing admixture used with a Portland cement concrete have been found.

Keywords: Contaminated land, Contaminants, Sulfate, Cubes, Guidance

Dr Stephen L Garvin is Head of Materials Technology & Performance at BRE Scottish Laboratory. His main research topics are concerned with building materials technology and performance, in particular the durability of building materials in aggressive environments. He is a Member of the Concrete Society (MCS) and the Royal Society of Chemistry (CChem MRSC).

Julian P Ridal is a scientist at BRE Scottish Laboratory. His main research topics are concerned with the performance of building materials and the impact of construction on the environment.

Max Halliwell is a senior scientist at BRE Garston. His main research interests are in the durability of concrete where he has investigated the effect of thaumasite sulfate attack on concrete durability.

INTRODUCTION

Contaminated land is a subject that has received increasing attention during the 1990s. These sites pose difficulties due to the presence of structures, uncharacterised fill material and contamination. A research project has been underway over a number of years that is intended to provide data on the durability of building materials used in contaminated land. If building materials can be shown to be durable in contaminated sites then the need for wholesale excavation and removal of soils can be avoided. At the same time any development of a contaminated site has to ensure that the risk to human health and the environment is acceptable.

Concrete can be chemically degraded by some contaminants. The potential for sulfate attack is well documented and guidance has been available for a number of years [1-3]. The ground conditions will also influence the potential for chemical attack. Whether concrete is attacked by a particular contaminant, therefore, depends on the following:

- the presence of water;
- the availability of the contaminant, its concentration and its replenishment rate;
- contact between the contaminant and the building material;
- the sensitivity of the material to the contaminant.

The following sections detail the experimental programme, results and the implications for guidance on the development of contaminated sites. Previous papers have also discussed the results that were obtained after one year's immersion in the contaminant solutions and slurry [4-7].

EXPERIMENTAL

The experimental programme has included four levels of test, as follows:

- accelerated laboratory tests involving the immersion of small mortar prisms in solutions of contaminants (duration of exposure 2-4 months);
- long term tests involving the immersion of 100 mm concrete cubes in solutions of contaminants (duration of exposure 5 years);
- long term exposure of 150 dia. x 300 mm concrete cylinders in a contaminated soil slurry (minimum duration of exposure 10 years);
- a field trial involving installation of concrete piles in a contaminated land site (minimum duration of exposure 10 years).

The results of small-scale mortar prism tests were reported [4] and previous papers have reported the methods used to assess deterioration of the concrete in the long-term tests [5-7]. Visual assessment, compressive strength measurements, chemical analysis, SEM and permeability have been used to assess deterioration. Table 1 shows the mixes that were selected for the experimental programme. Table 2 shows the solutions that were formulated in order to assess potential deterioration of different concrete mixes in contaminated soil.

Table 1 Details of concrete mixes used in test programme

MIX	CEMENT COMPOSITION	CEMENT RATIO	W/C RATIO (MAX)	CEMENT CONTENT (kg/m^3)	FINE AGG. (kg/m^3)	COARSE AGG. (kg/m^3)	28 DAY CS (N/mm^2)
1	PC	100	0.55	220	1200	910	34
2	PC	100	0.55	300	772	1192	59
3	PC	100	0.6	300	772	1192	42
4	PC/ggbs	30:70	0.57	320	769	1075	36.5
5	PC/ggbs	30:70	0.45	380	729	1055	52
6	PC/pfa	70:30	0.58	320	829	1025	38.5
7	PC/pfa	70:30	0.45	380	799	995	30
8	SRPC	100	0.45	360	710	1175	73
9	HAC/ggbs (BRECEM)	50:50	0.55	320	700	1175	21

PC - Portland cement; ggbs - ground granulated blastfurnace slag; pfa - pulverised fuel ash; SRPC - sulphate resisting Portland cement; HAC - high alumina cement. Mix 1 had a superplasticiser added to achieve adequate workability

Table 2 Solutions of contaminants used in concrete immersion tests

CONTAMINANT	SOLN A	SOLN B	SOLN C	SOLN D	SOLN E	SOLN F
pH	7.8	2.5	8	7.8	7	2.5
Sulphate	4200	4200	4200	4200	4200	4200
Sulphide	34	34	-	-	-	-
Chloride	-	-	218	24	-	-
Arsenic	5	5	65	1	-	-
Cadmium	-	-	-	0.8	-	-
Chromium	23	23	39	41	-	-
Cobalt	-	-	-	11	-	-
Copper	26.5	26.5	180	103	-	-
Lead	91	91	530	130	-	-
Manganese	355	355	-	730	-	-
Mercury	-	-	2	0.1	-	-
Nickel	23.5	23.5	28	43	-	-
Vanadium	-	-	-	47	-	-
Zinc	65	65	351	175	-	-
Phenol	1	1	3.5	-	-	-
Toluene	100	100	100	100	-	-

note - all concentrations in mg/l (except pH), water control solution termed soln G

RESULTS

Visual Assessment

A visual assessment was made of the condition of the concrete cubes after they were removed from solution at 3 years.

Portland Cement - Mixes 1, 2 and 3

The Portland cement mixes (2 and 3 - Table 1) were the most badly deteriorated of all the mixes. Mix 3 (of higher w/c ratio) had all but disappeared from solution C. Solution E (sulfate) had not caused as significant deterioration as the other solutions. In the acidified solutions (B and F) the degree of visual deterioration was not as great, but the acid left a brown and weathered appearance. Mix 1 of lower cement content (PC 220 kg/m^3), had deterioration. In most solutions the condition could still be regarded as at least moderate.

GGBS Concrete - Mixes 4 and 5

The weaker ggbs concrete (mix 4 - 320 kg/m^3) demonstrated more signs of deterioration than the stronger and more resistant concrete (mix 5 - 380 kg/m^3). Mix 4 had been deteriorated by all the solutions. In contrast with the PC concrete the acidified solutions had also caused deterioration of mix 4.

PFA Concrete - Mixes 6 and 7

The pfa concrete mixes were generally in good condition in all the solutions. The lower cement pfa concrete (mix 6) did show some signs of chemical reaction with the solutions. In particular in solutions A, C and D where expansion had occurred. The acidified solutions (B and F) had caused surface erosion of the cubes and some loss of material.

BRECEM and SRPC Concrete - Mixes 8 and 9

These concrete mixes were in good condition in all the solutions after three years. The main effect evident was the erosion of the surface layers in the acidified solutions for both mixes.

Compressive Strength

The compressive strength of cubes has been measured after three years immersion in the solutions. Control samples which have been cured in tap water since their manufacture are used for comparison to assess relative gain or loss of strength, see Table 3. In addition comparison of the 28 day compressive strength is given in Table 4.

PC Concrete - Mixes 1, 2 and 3

The cubes of mixes 2 and 3 that had been immersed in solutions A, C and D were too badly deteriorated to test for strength. Other samples could be capped with a mortar and tested. The control samples had not gained strength appreciably over the three years and therefore the results in Tables 3 and 4 are similar. The strength of mixes 2 and 3 after three years was about 20% of the control or 28 day value in solution B. For the two sulfate solutions (E and F), it was solution E with the higher pH that caused greater loss of strength. The lower pH solution, F, did not cause as much strength loss after three years. Mix 1 was typically about 70% to 75% of the 28 day value in the solutions, Table 4. Only solution B caused greater loss of strength at 60% of the 28 day value. Comparison with the control samples indicated that the strength was typically 60% to 64% of water stored samples.

Table 3 Compressive strength of concrete cubes as a percentage of control samples

	MIX 1	MIX 2	MIX 3	MIX 4	MIX 5	MIX 6	MIX 7	MIX 8	MIX 9
Sol A									
1 year	85.7	72.1	49.4	102.8	97.7	99.0	96.9	94.9	120.3
2 year	NT	21.5	15.0	74.7	NT	NT	NT	NT	NT
3 year	63.3	STD	STD	50.0	77.1	94.4	86.1	82.4	97.7
Sol B									
1 year	70.0	72.1	54.3	76.4	84.1	90.0	69.4	100.6	110.9
2 year	NT	27.4	22.2	52.9	NT	NT	NT	NT	NT
3 year	51.9	21.5	17.0	STD	66.7	43.9	33.7	68.2	70.1
Sol C									
1 year	88.6	96.4	72.8	83.0	100.0	101.0	94.9	103.8	103.1
2 year	NT	51.4	15.4	74.7	NT	NT	NT	NT	NT
3 year	63.3	STD	STD	86.4	84.0	94.4	103.0	89.2	98.9
Sol D									
1 year	95.7	76.6	50.6	99.1	97.7	100.0	98.0	104.5	95.3
2 year	NT	40.9	16.9	66.5	NT	NT	NT	NT	NT
3 year	63.3	STD	STD	64.4	86.1	100.0	99.0	91.5	101.1
Sol E									
1 year	95.7	88.3	75.3	98.1	100.0	113.0	87.8	108.3	115.6
2 year	NT	41.2	38.7	67.1	NT	NT	NT	NT	NT
3 year	60.8	21.5	21.6	63.6	88.9	94.4	98.0	85.2	94.3
Sol F									
1 year	90.0	91.9	79.0	84.0	86.4	95.0	67.4	99.4	112.5
2 year	NT	61.9	39.5	59.1	NT	NT	NT	NT	NT
3 year	59.5	38.0	35.2	50.0	63.2	59.8	56.4	71.6	71.3

NT = Samples not due to be tested after two years
STD = Samples too badly deteriorated to be tested
control samples have been cured in tap water for 3 years

GGBS Concrete - Mixes 4 and 5

Table 4 indicates that the ggbs concrete control samples, of mixes 4 and 5, have attained 160% and 140% of their 28 day value respectively. The samples which have been immersed in solutions while gaining strength compared to their 28 day value in some cases did not attain as high a strength as the control samples at all stages during immersion, Table 3. In most solutions the samples lost strength between one and three years.

PFA Concrete - Mixes 6 and 7

The control samples of pfa concrete attained 140% and 170% of their 28 day values after three years, Table 4. In most of the solutions the concrete had gained strength equivalent to the control samples. The exception being the acidified solutions which caused significant loss of strength of both these mixes by three years. Solution B with the low pH and mix of contaminants resulted in the most significant loss of strength.

SRPC Concrete - Mix 8

The initial 28 day strength of the SRPC concrete was high, but the control samples have continued to gain strength over three years, Table 4. In general the strength of the SRPC cubes in the neutral solutions (A, C, D and F) has remained at about the 28 day level. The exception being the acidified solutions (B and F) which have lost strength between one and three years.

BRECEM Concrete - Mix 9

The BRECEM concrete had a low 28 day compressive strength. However, there has generally been a large and consistent gain in strength to one and then three years. Comparable strength gains have been noted for the BRECEM concrete in near neutral conditions. Only acidified solutions have resulted in loss of strength from one to three years. There was an initial gain in strength to one year, but the strength at three years was about 150% of the 28 day value, Table 4.

Chemical Analysis

Chemical analysis has been carried out to determine the concentration of sulfate in the concrete cubes at various depths. At one year it was generally found that sulfate levels in the surface layers (1 - 6 mm) were higher than in the bulk of the material. At greater depth than 6 mm the sulfate levels were generally at background levels. The exception was where the solutions had been acidified and sulfate penetration had occurred to a greater extent, generally about 11 mm.

Table 4   Compressive strength of concrete cubes as a percentage of 28 day value

	MIX 1	MIX 2	MIX 3	MIX 4	MIX 5	MIX 6	MIX 7	MIX 8	MIX 9
Sol A									
1 year	88.2	67.8	47.4	148.7	125.7	128.6	160.2	101.6	184.8
2 year	NT	24.6	15.8	115.5	NT	NT	NT	NT	NT
3 year	73.5	STD	STD	80.5	108.1	131.2	146.7	98.9	204.0
Sol B									
1 year	72.0	67.8	52.2	110.5	108.1	116.9	114.6	107.8	170.4
2 year	NT	27.4	23.3	81.8	NT	NT	NT	NT	NT
3 year	60.3	22.0	17.8	STD	93.5	61.0	57.3	81.9	146.4
Sol C									
1 year	91.2	90.7	69.9	120.0	128.6	131.2	156.8	111.2	158.4
2 year	NT	51.4	16.2	115.5	NT	NT	NT	NT	NT
3 year	73.5	STD	STD	139.1	128.6	131.2	175.3	107.1	206.4
Sol D									
1 year	98.5	72.0	48.6	143.2	125.7	129.9	161.8	111.9	146.4
2 year	NT	40.9	17.8	102.7	NT	NT	NT	NT	NT
3 year	73.5	STD	STD	103.7	120.8	139.0	168.6	109.8	211.2
Sol E									
1 year	98.5	83.1	72.3	141.8	128.6	146.8	145.0	116.0	177.6
2 year	NT	41.2	40.7	103.6	NT	NT	NT	NT	NT
3 year	70.6	22.0	22.5	102.3	124.7	131.2	166.9	102.3	196.8
Sol F									
1 year	92.7	86.4	75.9	121.4	111.1	123.4	111.3	106.4	172.8
2 year	NT	61.5	41.5	91.4	NT	NT	NT	NT	NT
3 year	69.1	39.0	36.8	80.5	88.6	83.1	96.1	85.9	148.8
Sol G									
1 year	102.9	94.1	96.1	144.6	128.6	129.9	165.2	107.1	153.6
2 year	NT	100.0	105.0	154.6	NT	NT	NT	NT	NT
3 year	116.2	102.5	104.4	160.9	140.3	139.0	170.3	120.1	208.8

Sol G = water control;   NT = Samples not due to be tested after two years
STD = Samples too badly deteriorated to be tested

After three years a number of cubes, especially mixes 2, 3 and 4, could not be analysed as they were too badly deteriorated. The levels of sulfate in cubes subjected to near neutral pH solutions (A, C, D and E) displayed similar levels of sulfate to that at one year. The highest concentrations being in the surface layers and reducing concentration beyond 6 mm. In the acidified solutions (B and F) the level of sulfate was generally high through the cubes.

SEM Analysis

A number of samples, mixes 1, 2, 3, 4 and 6, were selected for SEM analysis. Samples were analysed for ettringite and thaumasite formation, and for the condition of calcium-silicate-hydrates (CSH).

Ettringite was found in a number of samples of mix 2, this was particularly for samples taken from the surface of the cube. Samples from within the cubes, taken at a depth of 50 mm, either showed no signs of ettringite or were lower than the surface concentration. The other mixes did not generally show ettringite formation apart from mix 4 (PC/ggbs) which showed abundant ettringite in the surface region when exposed to solution C. The CSH structure in this case contained a number of pores, all of which had become filled with ettringite.

DISCUSSION AND CONCLUSIONS

The performance of concrete in the various solutions is most easily indicated by reference to cement type, followed by cement content. The Portland cement concrete (mixes 1 - 3) had the poorest overall performance. In particular mixes 2 and 3 had deteriorated to the extent of near complete deterioration in a number of solutions. The acidified solutions (B and F) did not cause the same degree of deterioration by visual assessment as near neutral solutions, however, the strength was much reduced.

The reason for this considerable strength loss but without the obvious deterioration is concerned with the acid solution that has leached calcium hydroxide from the concrete. This has left a much more porous structure into which sulfates can penetrate. The resultant reaction between sulfate and the calcium-aluminate phases did not cause deterioration of the concrete as the ettringite formed had space in which to grow. However, there was a loss of cementing in the concrete and strength reduced substantially.

The surprising result with respect to the PC concrete was that of mix 1. This mix had the lowest cement content (220 kg/m^3) but performed better than the concrete with higher cement content (300 kg/m^3). The reason for the better performance was connected to the use of a superplasticiser to achieve adequate workability in the mix. This reduced the water demand for such a mix and allowed the specified maximum w/c ratio (0.55) to be achieved.

The performance of the pfa, SRPC, BRECEM and the higher cement content ggbs concrete was good in all the solutions. The exception being in acidified solutions where most of the concrete samples had lost about 20% strength from their 28 day value. The acidification of solutions had caused erosion of surface layers on all cubes and this resulted in exposure of aggregate. An unusual result of these more resistant concrete samples in the acidified solutions was that high levels of sulfate were found in the surface layers and this extended into the bulk of the cubes. This was in contrast to the SEM observations that indicated only minor sulfate presence, through observation of ettringite and other sulfate minerals. It is considered that the acidification caused increased porosity of the cubes. This has allowed sulfate to enter the pore system, but there is not the same amount of calcium-aluminates in these mixes with which to undergo chemical reaction. Therefore, the high sulfate concentration but low ettringite formation observed.

After one year, the results which were reported [5,6] indicated that the solutions of multiple contaminants caused enhanced deterioration of the concrete cubes in comparison with the cubes exposed only to sulfate. After three years these conclusions are still valid, but are most readily seen for concrete mixes that performed poorly. There was some evidence that the more resistant concrete mixes had their strength development up to one year retarded. Over the period one to three years the rate of strength gain, including the control samples, was low.

Ultimately, the question of the impact of these results on current guidance needs to be addressed and further guidance supplied. The research has not reached the stage of being able to produce new guidance. However, such guidance would most likely be based on revisions to BRE Digest 363 [1] and BRE Report BR255 [2]. The reason for including the revision in the Digest is that sulfates are still the dominant species for chemical attack. As well as concentrations of contaminants, other factors such as ground permeability and degree of replenishment of contaminants at the concrete surface are important. These variables take the guidance into the realms of risk assessment. This would best be considered with the revision of the BRE Report that is specific to contaminated land. The revision of the report would include a risk assessment methodology for concrete in contaminated land.

The experimental work is continuing and this paper has reported interim results after three years. The final results will be obtained and the revision of guidance will be completed in due course.

ACKNOWLEDGEMENTS

The authors would like to thank the Department of the Environment, Transport and the Regions for funding this research.

REFERENCES

1. BUILDING RESEARCH ESTABLISHMENT. Sulfate and acid resistance of concrete in the ground, BRE Digest 363, CRC Ltd, London, 1991 (with amendments 1996).

2. BUILDING RESEARCH ESTABLISHMENT. The performance of building materials in contaminated land, BRE Report BR255, CRC Ltd, London, 1994.

3. BICZOK, I. Concrete corrosion, concrete protection, Budapest, Hungarian Academy of Science, 1972.

4. PAUL, V, AND UBEROI, S. Determination of the resistance of cements using small-scale accelerated tests, Proceedings of the International Conference on the Implications of Ground Chemistry/Microbiology for Construction, University of Bristol, June, 1992.

5. GARVIN, S L, AND LEWRY A J. The performance of concrete in contaminated land, Conference on Concrete for Environment Enhancement and Protection, Dundee, June, 1996, pp457-469.

6. GARVIN, S L, AND LEWRY, A J. Construction on Contaminated Land, Proceedings of the COBRA '95 RICS Construction and Building Research Conference, Volume 1, Edinburgh, 1995, pp 203 - 214.

7. GARVIN, S L, LEWRY, A J, AND RIDAL, J P. Building materials performance in contaminated land : experimental and field work, Proceedings of the Third International Symposium on Environmental Contamination in Central and Eastern Europe, Warsaw, September, 1996, pp 835 - 837.

PROPOGATION OF CRACKS IN FIXED-FIXED REINFORCED CONCRETE BEAMS

M M Tawil

University of El-Fateh

Libya

ABSTRACT. Many normally reinforced concrete structures stiffer from cracking due to changes in environmental conditions as well as to alternations in loading conditions acting either separately or in a combination. Cracking of a concrete element will reduce its stiffness to some extent that means changing the deformation of the whole member as well as the distribution of bending moments along statically indeterminate structures. Redistribution of bending moments along an indeterminate member will depend mainly on its boundary conditions and type of loading subjected to which define the regions and extent where cracks do propagate. In this paper the propagation of cracks in fixed-fixed beam, loaded once with a concentrated point load at mid span and the other case with a uniformly distributed load on the whole span, has been studied and presented. Results for the effect of these cracks on deflection and bending moment redistribution are given for a number of ratios relating the moment of inertia of cracked section to the moment of inertia of uncracked section, and proper factors taking this effect into consideration when designing of such beams have been suggested.

Keywords: Cracking, Moment redistribution, Deformation, Statically indeterminate structures, Stiffness, Equilibrium, Boundary conditions.

Professor Mustafa M Tawil is a professor of civil engineering, University of El-Fateh, Tripoli, Libya. He specializes in analysis and design of concrete structures, with particular reference to the numerical technique of finite integral method. Professor Tawil has contributed more than forty five technical end professional papers in the last fifteen years which are published in symposium and conference proceedings, local professional magazines and regional and international periodicals. He is a member of quite a number of technical and professional committees. Professor Tawil is a member of board of directors of National Consulting Bureau of Tripoli, which is a leading firm in the country, for the last ten years and he has been also a part time consultant to a number of other firms and organizations.

INTRODUCTION

Cracking in structural elements made of normal reinforced concrete is generally unavoidable sometime during its life span. Cracked sections in an element of a structure will affect the overall stiffness of the system because of the reduction in their moments of inertia thus changing the behavior of the structural system as a whole. This reduction in overall stiffness will increase the deformations of the cracked elements in general [1], and will change the behavior of the structural system resisting the applied loads internally when it is of statically indeterminate type. The bending moment redistribution in cracked propped cantilevers under two cases of external loads applied symmetrically has been studied by the author [2]. In this paper the case of fixed-fixed beams is considered where again two cases of externally applied loads are covered; that is a concentrated point load acting at mid span of the beam and a uniformly distributed load acting on the whole span. Study of this phenomenon will cover a practical range of ratios relating the moment of inertia of a cracked section to the moment of inertia of an uncracked section [3], and considering that all cracked sections as being fully cracked. The method of conjugate beam has been used for calculating the fixedend moment and the maximum deflection at mid span for these beams.

The aim of this study is to try to reach defining the factors that may be applied on existing formulas, given by some international codes of practice, for maximum bending moments that are used in analysis and design of such type of beams in order to take their expected future increase due to cracking effect into consideration. This modification of maximum moments will serve the purpose of keeping the designed concrete sections not being underdesigned when cracking occurs.

CRACKING PHENOMENON IN NORMAL RC STRUCTURES

Once the tensile strength in a normal reinforced concrete element exceeds its modulus of rupture cracking will appear if not prevented by a suitable amount of reinforcement. The bending moment at which cracks initiate is called the cracking moment, and if this moment increases for any reason crack will propagate around that initial point causing regions with reduced stiffness which will increase the size of deformation of sections along the structural member, and in the case of statically indeterminate structures will also redistribute the values of bending moments along the span to maintain equilibrium of such structural system.

CRACK PROPAGATION IN RC FIXED-FIXED BEAMS

Concentrated Point Load at Mid Span

When a fixed-fixed prismatic member of length (L) is loaded with a concentrated load (P) acting at mid span the absolute magnitude of both maximum positive and negative bending moments is the same and equal to (PL/8). Theoretically speaking the initiation of cracks will start simultaneously at both ends and at mid span ofthe beam when those moments reach the value of cracking moment and cracking will continue to propagate if the bending moments are increased due to any reason forming then three regions of cracked sections as shown in Figure 1 two regions at the ends and one region in the middle.

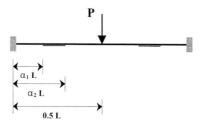

Figure 1 Cracking regions for concentrated load at mid span

Uniformly Distributed Load

In this case the bending moments at the fixed supports are larger in magnitude than that at the mid span. The cracks when initiated will start at the fixed supports first leading to two regions of cracking, Figure 2a. The magnitude of moments at those fixed supports will be then reduced due to this cracking and the moment at mid span will start to increase accordingly to satisfy equilibrium conditions. If cracks continue to propagate, due to any reason, until the magnitude of the moment at mid span reaches the value of cracking moment then a third region of cracking will appear, Figure 2b.

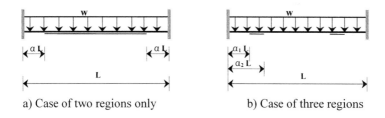

a) Case of two regions only   b) Case of three regions

Figure 2 Cracking regions for uniformly distributed load

MOMENT REDISTRIBUTION DUE TO CRACKING

Concentrated Point Load at Mid Span

Due to the fact that the magnitude of maximum positive and maximum negative bending moments for this kind of load is the same, the factors (α_1 and α_2, defining the end of the first cracking region and the start of the second cracking region respectively as shown in Figure 1, should add up to 0.5 at all times.

In this case no bending moment redistribution is performed for all values of ratio of α_1, being less than or equal to (0.25) and for all values of β where β is defined as:

$$\beta = I_{cr} / I_g$$

where, I_{cr} = moment of inertia of cracked section

I_g = moment of inertia of uncracked section

Uniformly Distributed Load

Figure 3 shows the percentage increase in the value of maximum positive bending moment at mid span of the fixed-fixed beam due to cracking of concrete and for values of β of (0.2, 0.4, 0.6 and 0.8). The value of this moment with no cracking is ($wL^2 / 24$), where (w) is the intensity of the applied uniformly distributed load.

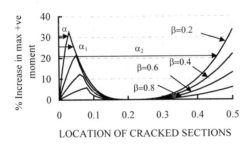

Figure 3 Percentage increase in max. +ve moment at mid span

The moment redistribution along the whole fixed-fixed beam for $\beta = 0.2$ is shown in Figure 4 at the instant when the positive moment at mid span is absolute maximum. It should be noted that the absolute maximum positive moment is reached when cracking is just initiated in a section at mid span of the beam.

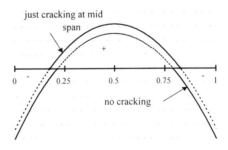

Figure 4 Moment redistribution when moment at mid span is absolute maximum ($\beta = 0.2$)

CRACKING EFFECT ON DEFLECTION

Concentrated Point Load at Mid Span

Although it was stated above that no moment redistribution is taking place under this kind of load, it is to be noted that the effect of cracking on displacement is appreciable. Figure 5 shows the magnification factor (γ) of the original maximum deflection at mid span against cracking length factor (α) and for different values of β The value of the maximum deflection δ_{max} is given as:

$$\delta_{max} = \frac{PL^3}{192\,EI_g}(\gamma)$$

where, E = Young's modulus of elasticity of concrete

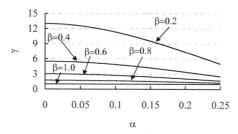

Figure 5 Magnification factors for max. deflection at mid span

Uniformly Distributed Load

Figure 6 shows the magnification factors to be used in calculating the maximum deflection at mid span for this kind of loading. This deflection is expressed as:

$$\delta_{max} = \frac{wL^3}{384\,EI_g}(\gamma)$$

DISCUSSION OF RESULTS

It has been shown that there will be no moment redistribution due to cracking in fixed-fixed beams under concentrated point load acting at mid span. In case of uniformly distributed load, however, the moment redistribution due to cracking in concrete will happen and the maximum increase in maximum positive moment at mid span can tee calculated from the following suggested relation:

$$\text{increase} = (1 - \beta)(48 - 34\,\beta + 15\,\beta^2)$$

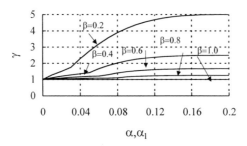

Figure 6 Magnification factors for max. deflection at mid span

This suggested relation is shown graphically in Figure 7.

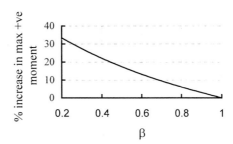

Figure 7 Percentage increase in maximum +ve bending moment
due to cracking (uniformly distributed load)

The increase in maximum deflection at mid span can be estimated from Figure 5 for the case of concentrated point load at mid span of the fixed-fixed beam and from Figure 6 for the case of uniformly distributed load.

CONCLUSIONS AND RECOMMENDATIONS

The effects of cracking in concrete on the moment redistribution and displacements for fixed-fixed beams loaded with two different symmetrical loading systems have been highlighted. The maximum positive moment in the case of point load at mid span is found not to change during cracking due to the fact that cracking will initiate simultaneously both at fixed ends and at mid span.

In the case of uniformly distributed load the moment redistribution is studied and a relation has been suggested to modify the value of maximum positive bending moment at mid span for fixed-fixed beams under such loading in order not to underestimate the actual value of moment when cracking occurs. Two different curves related to effect on maximum deflection at mid span are given and can be used to estimate such values for different values of ratios relating the moment of inertia of cracked sections to the moment of inertia of uncracked section.

It is recommended that other types of loading be studied for this type of structural system and similar studies be performed to generalize the results of the effect of cracking in normal reinforced concrete sections. The purpose is to maintain the cracked structures standing resisting the applied loads safely even though their stiffness values have been reduced.

REFERENCES

1. TAWIL, MUSTAFA, M.   Stiffness of cracked rectangular reinforced concrete beams, Proceedings of the 4th Int Conference on Concrete Technology in Dweveloping Countries, Gazimagusa, North Cyprus, 1996, pp 441-448

2. TAWIL, MUSTAFA, M.   Cracking effect on moment redistribution in RC propped cantilevers, Proceedings of the 5th Int Conf on Deterioration and Repair of Reinforced Concrete in the Arabian Gulf, Vol 1, Manama, Bahrain, 1997, pp61-69

3. WANG, C K, SALMON, C G. Reinforced concrete design, 2nd Edition, Intext educational publishers, 1973,

A STUDY OF THE SECONDARY FLEXURE IN UNIAXIAL TENSILE TEST OF CONCRETE

H Akita

H Koide   M Tomon

Tohoku Institute of Technology

K Hosokawa

Mitsubishi Materials Co Ltd

Japan

ABSTRACT. There are three difficulties associated with carrying out uniaxial tensile tests. These are prevention of an unstable fracture, a secondary flexure and duplicate cracks. To overcome these problems, a novel technique, which consists of applying notches and attaching an adjusting gear system, is examined in this paper. Applying notches to test specimens reduced the tendency of decreasing tensile strength despite generation of high stress concentration. Consequently, it is possible to yield a stable fracture and prevent a secondary flexure by using this technique. Additionally, the effect of the duplicate cracks was diminished. Therefore, this technique is suggested as a standard test procedure for determining the uniaxial tensile strength of concrete.

Keywords: Uniaxial tension, Test method, Secondary flexure, Concrete, Softening

Hiroshi Akita is a Professor in the Department of Civil Engineering at Tohoku Institute of Technology. His research interest relates to the fracture mechanics of concrete and the water movement within cementitious materials.

Hideo Koide is an Associate Professor in the Department of Civil Engineering at Tohoku Institute of Technology. His research interest relates to the fracture mechanics of concrete and the reliability analysis of concrete structures.

Masanao Tomon is a Professor in the Department of Civil Engineering at Tohoku Institute of Technology. His research interest relates to the durability of concrete.

Kiyoshi Hosokawa is a Chief Research Worker in Concrete Center of Mitsubishi Materials Co. Ltd. His research interest relates to the design and application of high performance concrete.

INTRODUCTION

The knowledge of tension softening process of concrete is beneficial to understand the fracture mechanism, to analyze the fracture behaviour, and further to improve properties of concrete. The best way to investigate the tension softening process is applying uniaxial tension force directly on an unnotched concrete specimen, if it is possible, because it can provide more reliable information than any other alternative methods. However, there are several problems to perform the uniaxial tension test, such as an unstable fracture, a secondary flexure and duplicate cracks. To maintain a stable fracture throughout concrete during the test is required for monitoring post-peak loading behaviour in detail. The secondary flexure, which is usually generated by tension softening at the weakest zone of the specimen, should be prevented because it reduces the observed peak load. However, when the secondary flexure is prevented, another problem can appear, i.e. duplicate crack formation.

A novel technique, which will be introduced in this paper, is able to solve these problems for performing the uniaxial tensile test of concrete. This technique makes it possible to yield a stable fracture as well as to prevent a secondary flexure during testing. Also, the method to minimize the formation of the duplicate cracks will be addressed. Therefore, this technique has a potential to utilize as a standard test for the uniaxial tensile test of concrete.

ESTABLISHING STABLE FRACTURE

For performing uniaxial tension test, it is essential to establish a stable fracture during testing. Patersson explained that the characteristic length is almost identical to the limit length of a stable fracture regardless of the rigidity of a loading machine [1]. The entire length of a concrete specimen should be identical to the characteristic length under load-controlled or displacement-controlled testing because these two methods count total elongation of the specimen. However, as shown in Table 1, the characteristic length decreases as compressive strength increases, thus these methods are limited to explore low strength concrete or a relatively small specimen. Even worse, it is too complicated to prevent the secondary flexure.

Table 1 Examples of characteristic length and related data,
related data were referred to Kitsutaka et al. [2]

Compressive strength	f_c (N/mm²)	30	60	100
Tensile strength	f_t (N/mm²)	2.2	4.0	5.0
Young's modulus	E (GPa)	25	32	38
Fracture energy	G_r (N/m)	100	110	125
Characteristic length	L_{ch} (cm)	52	22	19

When the applied load is controlled by strain (or deformation), the characteristic length should not be the same as the entire specimen length but as a strain gauge length (or LVDT length). In other words, the entire specimen is independent of the characteristic length. Thus, the latter two methods enable to investigate higher strength concrete with bigger size. Therefore, the strain-controlled method was employed in this study.

PREVENTING SECONDARY FLEXURE

The uniaxial tension tests focused on the tension-softening behaviour have been discussed for the past 30 years. However, only a few investigations have carefully considered the effect of a secondary flexure on the tensile behaviour. Since the secondary flexure reduces the peak load and consequently the tensile strength, it should be effectively prevented. Van Mier et al. tried to prevent the secondary flexure by fixing platens at both ends of a specimen [3]. Figure 1 shows their stress-deformation diagram.

The numbers of 1-4 indicate the employed LVDTs at four positions along the circumstance of the cylindrical specimen. Since the average axial deformation (6 in diagram) was used as a feedback signal for loading, they concluded that the effect of the secondary flexure might be negligible. However, the effect of the flexure seems to be greater than expected. The strain distribution at the maximum stress of Figure 1 was schematically illustrated in Figure 2 (a). The deformation of LVDT 2 at the maximum stress is roughly one third of that of LVDT 4. It indicates that the deformation of LVDT 4 is thought to include a crack opening displacement.

Figure 2 (b) shows the calculated stress distribution from the strain distribution (Figure 2 (a)), assuming that elastic behaviour governs stress-strain relationship until ε_t, while the tension softening affects the relationship after ε_t. If the secondary flexure were completely prevented, the stress should be distributed evenly at f_t like Figure 2 (c). However, the stress distribution in van Mier et al.'s experiment is uneven, as shown in Figure 2 (b), and gives about 10% smaller peak load than exact one. The result comes from the incomplete prevention of the secondary flexure. Therefore, the fixed platen method is thought to be insufficient to prevent the secondary flexure.

Employing additional actuators is useful to avoid the secondary flexure, as Heilmann reported [4]. However, duplicate cracks will be initiated inevitably, if the secondary flexure were completely prevented as van Mier reported [5]. Since the existence of the duplicate cracks misleads tensile behaviour, it should be also avoided. The duplicate crack forms more easily on an unnotched or too long LVDT length specimen.

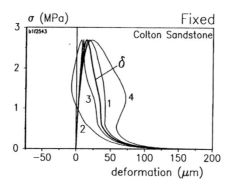

Figure 1 Stress deformation diagram by van Mier et al

Carpinteri et al. innovated test instrument including actuators [6], resulting the first successful prevention of the secondary flexure among researchers, as shown in Figure 4. Little deviation of the profile of each position from the average profile indicates that the stress of all critical section may reach the maximum value (f_t in Figure 2 (c)). Using a specifically designed specimen, shown in Figure 5, made it possible to avoid duplicate cracks. However, the complicated shape of a specimen generates a complex stress distribution vertically (Figure 5 (a)) and horizontally (Figure 5 (b)). Their procedure has some disadvantages in high equipment cost as well as too complicated specimen fabrication.

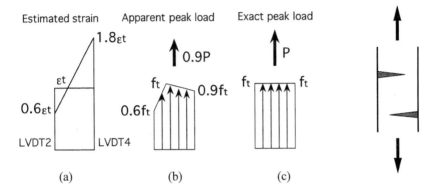

Figure 2 Error of peak load

Figure 3 Duplicate cracks

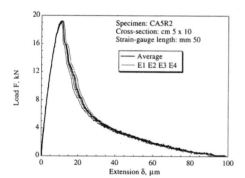

Figure 4 Load extension diagram by
Carpinteri et al

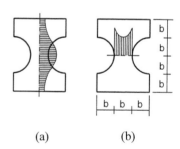

Figure 5 Tensile test specimens
of Carpenteri et al

REEVALUATION OF NOTCH

Notched specimens have been used by many investigators for two reasons. One is to produce a stable fracture and the other is to enable observation and control by fixing the crack location. However, notches also generate high stress concentration and considered to mislead the observation of the tensile strength.

Therefore, the authors had used unnotched specimens for these 4 years [7]. Recently, they found in simulation results that notches produce only a little reduction of tensile strength in spite of generating a high stress concentration. A brief explanation is the following, and details are described in elsewhere [8].

Figure 6 shows a schematic explanation of stress distributions on notched section of a concrete specimen as stress increases. The stress near a notch tip already reaches the ultimate tensile strength due to the high stress concentration even at a low stress at the center (Figure 6(a)). As increase the stress level at the center, two softening zones proceed from both notch tips to the center (Figure 6 (b)). Since the crack opening displacement (COD) on the softening zone are very small, the descending of the cohesive stress there from ultimate tensile strength is also very small.

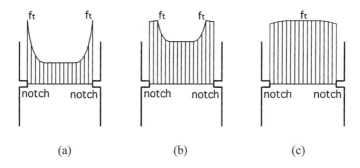

Figure 6 Stress distribution of notched cross section

As shown in Figure 6 (c), when the stress at the center reaches the ultimate tensile strength, almost a uniform distribution of the stress on the whole ligament cross section can be obtained. It is the most important reason why the authors reevaluate notched specimens. In addition, all the problems previously mentioned are solved by the adoption of notches.

EXPERIMENTAL DETAILS

Ordinary concrete (cement: sand: aggregate = 1: 2: 3.8, W/C=0.5) was used for investigating uniaxial tension test. The compressive strength and the tensile strength (by the Brazilian test) were 29.8 N/mm² and 2.57 N/mm², respectively. The geometry of the specimen was a prism of 100x 100x 400 mm. The middle of the side faces (except for cast and bottom faces) was cut by a diamond saw to make a 10-mm depth notch. A cross type steel frame was attached to each specimen end by epoxy resin and embedded bolt. A 70-mm extensometer was attached in the middle of each side face.

To prevent a secondary flexure, an adjusting gear system was attached, as shown in Figure 7. As the horizontal gear has 50 threads, the strain of 1x 10-6 on the specimen surface can be easily adjusted by turning the vertical gear by one tenth.

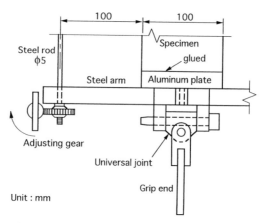

Figure 7 Adjusting gear system by the authors

The steel rod sustains only tensile forces and the maximum force was less than 2% of machine load, found in previous experiments. By turning the adjusting gear, a bending moment is given to the specimen, i.e. contracting the faced side and elongating the opposite side. It also helps to adjust load eccentricity due to gluing the steel frame to the specimen. Details of the gear system is described elsewhere [8].

The application of tensile loading was controlled by four sequential control stages, as indicated in Table 2. The testing began with the load control stage until the tensile force reached 15 kN, then shifted to the strain control stage. During strain control stages, the applied load was controlled by the average strain of the bottom face (ch 1) and the cast face (ch 3). The strain values of two notched sides (ch 2 and ch 4) were only used as an indication for preventing secondary flexure.

Table 2 Machine control stages for tensile loading

STAGE	CONTROL	DURATION	RATE
1	Load	15kN	50N/s
2	Strain	200×10^{-6}	0.1×10^{-6}/s
3	Strain	400×10^{-6}	0.2×10^{-6}/s
4	Strain	Last	0.3×10^{-6}/s

EXPERIMENTAL RESULTS

Figure 8 shows a typical load-deformation curve obtained by the present procedure. The load increased elastically until reaching the maximum of 20.5 kN, then decreased very rapidly to 15 kN followed by a gradual decrease until breaking point.

The deformation and the load at the breaking point were 0.124 mm and about 2 kN, respectively. It is thought that this test procedure monitored the whole range of the tensile softening behaviour for concrete successfully, because of the observation on a rapid decrease in the load after the peak and a long tail in the softening regime.

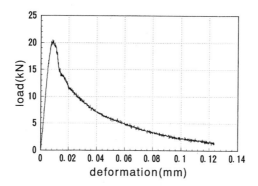

Figure 8 Load deformation curve

It was assumed that the outside of the softening zone still behaves elastically even after the maximum load. So, crack opening displacement (COD) was obtained by the following equation:

$$w = \delta - \frac{PL}{EA} - \delta_r$$

where δ: observed deformation, P: load, L: gauge length, A: cross sectional area of ligament, E: Young's modulus, and δ_r: residual elongation when the load decreases to zero. The schematic explanation for calculating COD is shown in Figure 9. The previous load-deformation curve was converted to tension softening curve, i.e. stress-COD curve, as shown in Figure 10. The total fracture energy of this specimen was estimated about 95.0 N/m from the total area under the curve with a linear extrapolation to the zero loads.

Figure 11 shows a history of the displacement of the platen with the observed deformation in the gauge length of two faces, i.e. cast (ch 1) and bottom (ch 3) faces, with respect to testing time. The platen displacement behaved a very rapid increase at the beginning of the loading, then followed by a gradual decrease after the peak. However, the deformation of gauges on both faces always increased. It means that the gauge zone is continuously elongated after the peak until breaking, being contrary to shrinking the total length due to the load drop. It indicates that such stable fracture cannot be obtained by displacement-controlled loading with the present specimens. Since the monitored deformations at both gauges were almost coincided, the secondary flexure was prevented effectively by the present method. The same result also obtained for notched face deformation. It was also observed that increasing a strain rate (see Table 2) allow the deformation rate of the gauge region greater.

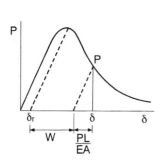

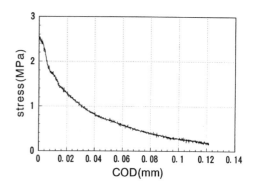

Figure 9 Schematic explanation
of calculating COD

Figure 10 Tension softening curve

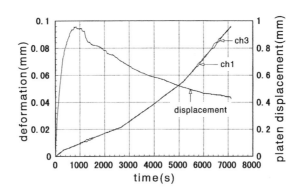

Figure 11 Histories of deformation and platen displacement

Since the deformation at the cast face (ch 3) usually tends to be greater than that at the bottom face (ch 1) during tension loading, as van Mier reported [5], the side of the ch 3 needs to be tightened to prevent the secondary flexure.

Figure 12 shows a history of the additionally applied tensile force by the adjusting gear system at the cast face (ch 3) side, transferring a bending moment to the specimen. Since the additional tensile force at the side only 2.5% in maximum of the machine load during the test, the influence of this additional force on the tension softening process may be negligible. Thus, this test procedure has a potential to prevent the generation of a secondary flexure completely by a simple adjustment.

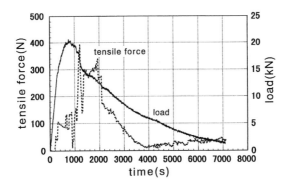

Figure 12 Histories of machine load and tensile force

As mentioned earlier, the duplicate crack formation becomes a serious problem when the secondary flexure is completely prevented. To reduce the tendency of the duplicate crack formation, Carpinteri used a complicated shape of concrete specimen [6]. Recently, the authors found in simulation that applying a notch on a specimen can also prevent the duplicate crack formation easily. Experimentally, no duplicate crack was observed when a specimen had a notch. Thus, the effects of the duplicate cracks can be ignored in this study.

CONCLUSIONS

This test procedure enables to monitor the entire range of tensile softening behaviour. It was possible to prevent the secondary flexure as well as to establish stable fracture. By applying a notch, duplicate cracks were effectively prevented. This test procedure has a potential to be a standard test method for investigation of uniaxial tension test because it has several advantages over other methods, e.g. easy to operate, use of simple shape specimen, and inexpensive equipment.

REFERENCES

1. LUND INSTITUTE OF TECHNOLOGY. Crack growth and development of fracture zones in plain concrete and similar materials. Reported by P E Patersson, Dec., 1981. 174pp.

2. KITSUTAKA, Y, NAKAMURA, S and MIHASHI, H. Simple evaluation method for the bilinear type tension softening constitutive law of concrete, Proc. Japan Concrete Institute, Vol.20, No.3, 1998, pp.181-186.(in Japanese)

3. VAN MIER, J G M, SCHLANGEN, E and VERVUUT, A. Tensile cracking in concrete and sandstone: Part 2 - Effect of boundary rotations, Materials and Structures, Vol.29, 1996, pp.87-96.

4. DELFT UNIVERSITY OF TECHNOLOGY. Deformation-controlled uniaxial tensile test on concrete, Reported by I D A Hordijk, Dec., 1989, 118pp.

5. VAN MIER, J G M and NOORU-MOHAMED, M B. Geometrical and structural aspects of concrete fracture, Engineering Fracture Mechanics, Vol.35, No.4/5, 1990, pp.617-628.

6. CARPINTERI, A and FERRO, G. Size effects on tensile fracture properties: A unified explanation based on disorder and fractality of concrete microstructure, Materials and Structures, Vol.27, 1994, pp.563-571.

7. KOIDE, H, AKITA, H and TOMON, M. A direct tension tests for obtaining tension softening curves of unnotched concrete specimens, Proc. Int. Symp. Brittle Matrix Composites, No.5, 1997, pp.366-375.

8. AKITA, H, KOIDE, H, TOMON, M and Sohn, D. A practical method for uniaxial tensile test of concrete, Concrete Science and Engineering, (in preparation)

EXPERIMENTAL INVESTIGATION ON FIBRE PULL-OUT FROM NORMAL AND HIGH STRENGTH CONCRETE

G Zingone

M Papia

G Campione

University of Palermo

Italy

ABSTRACT. Experimental results showing the behaviour of single fibres pulled out from concrete matrices are presented. Carbon, hooked steel and polyolefin fibres embedded in normal and high-strength concrete matrices are considered in order to simulate the actual physical condition occurring for a single fibre in the cracking stage of a fibre-reinforced concrete element. In this phase the presence of fibres ensures a bridging action on the cracks depending on the physical and mechanical properties of the components of the composite cementitious material, and the effects of these properties were investigated. Moreover, in order to take into account the random distribution and orientation of the fibres themselves in the matrices, different tests were carried out with variation in the inclination of the fibre with respect to the direction of the external load, as well as the embedded length. The results presented here give useful information on fibre-matrix interaction phenomena, on pull-out forces and slippages, and on the energy accumulated during the pull-out process.

Keywords: Composite materials, Fibre reinforced concrete, Normal and high strength concrete, Steel, Carbon and polyolefin fibres, Pull-out curves, Interface properties.

Professor Gaetano Zingone is a Professor of Structural Engineering at the Department of Structural and Geotechnical Engineering of the University of Palermo, Italy. He is Ph.D. coordinator at the same University, and his interest in research is on seismic engineering, structural problems in historic-monumental buildings, and nonlinear response of structures.

Professor Maurizio Papia is a Professor of Structural Engineering and Head of the Department of Structural and Geotechnical Engineering of the University of Palermo. His interest in research is on non linear analysis of structures, cyclic behavior of steel, reinforced concrete and masonry structures.

Dr Giuseppe Campione is a Ph.D at the Department of Structural and Geotechincal Engineeringat the University of Palermo, Italy. The experimental investigation described in this work was partially carried out while he was a visiting research engineer at the Department of Civil Engineering at the University of British Columbia, Vancouver, BC, Canada.

INTRODUCTION

In order to understand the mechanical behaviour of a fibre reinforced concrete, obtained by adding short randomly discontinuous fibres in a cementitious material, knowledge of the fibre-matrix interaction phenomena is required [1-2]. When the uniaxial tensile stress, induced in a non conventional reinforced member by external loads, reaches the tensile strength of the matrix, brittle failure arises, partly depending on the size of the member. In reinforced concrete members the cracking process of concrete can also determine a loss of bond in longitudinal bars or brittle shear failure, if inadequate anchorage length and transverse reinforcement are designed. The adding of randomly discontinuous short fibres in an appropriate percentage and length reduces the negative effects mentioned, because the increase in the tensile strength of the composite, due to the fibre-matrix interaction, ensures a bridging action of the fibres across the cracks in the case of both uncracked and cracked stages. Elastic interaction, debonding, frictional resistance and elongation of the fibres ensure ductile behaviour by a composite material, avoiding brittle behaviour, especially in the case of high strength concrete [3].

Recent studies [3-5] have shown that the behaviour of a fibre reinforced concrete, and especially its strength and energy capacity absorption, depends on the characteristics of the fibres, the cementitious materials and the interface properties between fibre and matrix. The possible use of different lengths and shapes of fibres randomly distributed in the matrices does not allow one to represent the actual pull-out process of single fibres with a simple model. Therefore, the pull-out response of single fibres is usually determined by experimental tests, varying the fundamental parameters of the fibres (material, length, diameter, shape, inclination angle with respect to external load), of the matrices (composition, strength, strain capacity) and of the interface between fibre and matrix.

In the present investigation the pull-out response of single carbon fibres was determined by varying the embedded length, the inclination angle of the fibre with respect to the external load and the strength of the matrices. The results are compared with those obtained by pulling out single aligned hooked steel and polyolefin fibres from high strength matrices. This particular interest in the experimental investigation of the pull-out response of single carbon fibres is related to the advantages which they offer with respect to steel and polyolefin fibres, like non- corrodibility and chemical and physical stability.

EXPERIMENTAL INVESTIGATION

Material Properties and Parameters Investigated

The pull-out tests of single fibres from cementitious matrices were carried out considering one series of six specimens for each variable investigated. With reference to carbon fibres having a rectangular transversal section of 0.2x2.39 mm and length varying between 20 and 30 mm, the influence of the following parameters was investigated: resistance of the cementitious matrix; inclination of the fibre with respect to the direction of the external load, with angles ranging from 0° to 90 °; length embedded in the matrix, considering fibres of 20 and 30 mm with embedded length of 10 and 15 mm, respectively.

Two matrix types were made, one with normal strength (48 N/mm²) and the other with high resistance (70 N/mm²). The characteristics of the matrices investigated are given in Table 1.

Table 1 Characteristics of the matrices tested

TYPE OF MATRIX	CEMENT (kg/m³)	WATER (kg/m³)	COARSE AGGREGATE (kg/m³)	SAND (kg/m³)	SILICA FUME (kg/m³)	SUPERPLAS-TICIZER (kg/m³)
NSC	300	165	1050	850	-	-
HSC	400	150	1050	720	55	8

NSC= normal strength concrete; HSC=high strength concrete

The characteristics of the fibres investigated are given in Table 2. This table includes the equivalent diameter ϕ and the length L_f of the fibres, the elastic modulus in tension E and the weight of the unit volume γ of the material constituting the fibre.

Table 2 Characteristics of the fibres

TYPE OF FIBRE	SHAPE	ϕ (mm)	L_f (mm)	f_t (N/mm²)	E (N/mm²)	γ (Kg/m³)
Poyolefin	——	0.80	25	375	12000	900
Steel	⌐___/‾	0.80	30	1115	207000	7860
Carbon	▬▬▬	0.78	20-30	800	100000	1340

Specimens and Experimental Set-up

Each specimen tested consisted of two concrete cylinders of height 63.5 mm and diameter 63.5 mm, superimposed and separated by a plastic sheet, as shown in Figure 1. The same fibre was anchored for half its length in each of the two cylinders, and inclined at an angle α with respect to the vertical direction, as shown in Figure 2.

The specimens were made in distinct phases. In the first phase the lower cylinder was made and the fibre embedded for a length of $L_f/2$ and inclined at an angle α with respect to the vertical between 0° and 90°. In the second phase, after twenty-four hours, the upper cylinder was made. The separating sheet was perforated in the zone where the fibre passed through and the hole was calibrated on the diameter of the fibre to avoid contact zones between the separate concrete casts. Each specimen was cured in a tank full of water saturated with lime inside a chamber with controlled humidity and temperature for twenty-eight days and then left in the open air for about an hour before the test. The pull-out tests were carried out using the equipment shown in Figure 3. For each test two steel rings of appropriate size were applied in the cast phase, one on the lower extremity of the lower cylinder, and one at the upper end of the upper cylinder.

Figure 1 Specimens used for pull-out tests on single fibres

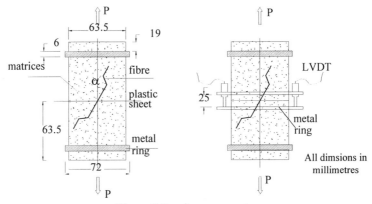

Figure 2 Specimen geometry

Figure 3 Test set-up

The rings made it possible to connect the specimens to the grips of the testing machine. The apparatus was inserted in a universal testing machine of the Instron type with load capacity of 150 KN equipped with a 3000 N load cell. The use of a load cell with a low scale makes it possible to record all significant values in a continuous and stable way.

The test was performed by pulling the two extremities of the cylinders clamped in the machine and automatically assigning a displacement gradient equal to 0.5 mm /min. The load cell placed between the machine and the specimen made it possible to measure the reactive force corresponding to each displacement value. The displacements were measured through two displacement transducers connected to the specimen by means of cylindrical rings (Figure 2) and placed symmetrically with 25 mm gauges. The load and the displacement were recorded by a digital acquisition system connected to the machine, operating at 16 bits with a frequency of 10 Hz. The electric signals were automatically transmitted to a computer which recorded the load and the displacement values by means of a data handling program.

EXPERIMENTAL RESULTS

Pull-out Tests on Single Carbon Fibres

Effect of embedded length and strength of matrices

Figure 4 shows results of pull-out tests on carbon fibres, parallel to the direction of the external load ($\alpha=0$) and of length 20 and 30 mm, from normal resistance (48 N/mm^2) and high resistance (70 N/mm^2) matrices. In both cases one can observe that with an increase in fibre length the maximum pull-out load increases. In all cases examined the ultimate state occurs due to failure of the fibre-matrix interface and never to excess of fibre strain. Moreover, the collapse value proves to be very far from the value for failure due to excess of tension in fibre. Comparison between the curves in Figure 4 shows that in the case of high-strength matrices the maximum pull-out load is 40 % higher, with the same fibre length, than the load for normal-strength concrete.

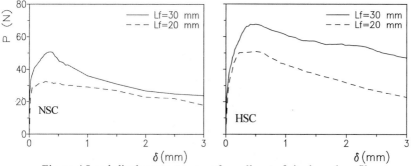

Figure 4 Load-displacement curves for pull-out of single carbon fibres
(L_f=20 and 30 mm) from NSC and HSC

In the case of matrices with high resistance in the descending branches of the curves, there is a more marked beneficial effect on the response, due to the length of the fibres, than in the case of normal strength concrete. This is probably due to the silica fumes present in high-strength matrices; with a more resistant interface there is also a more ductile response. Although the number of lengths considered is insufficient to allow one to draw definitive conclusions, it appears reasonable to assume a linear dependence of the maximum pull-out load on the length of the fibre immersed in the matrix. It appears evident that the limit state of the complex is reached on account of pull-out of the fibre due to the yielding of the interface while the ratio between the tangential stress at the elastic limit and the maximum debonding stress is greater than unity.

Effect of fibre inclination angle

Figure 5 shows the results relating to pull-out tests on single fibres of length 20 mm from high-strength matrices, with variations in the inclination of the fibre with respect to the direction of the applied load, respectively for angles of 0.00°, 22.50°, 45.00° and 67.50°. It can easily be observed that with increases in the inclination of the fibres the maximum pull-out load decreases, with a reduction ranging from 20 to 60 % with respect to the case of fibres aligned with the external load.

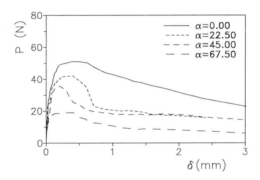

Figure 5 Effect of the inclination angle on the pull-out of single carbon fibres from HSC

This phenomenon probably has to be ascribed to the fact that when the fibre is inclined with respect to the direction of the external load local tensile stresses are triggered off inside the matrices, at the fibre-matrix interface. These produce local failure phenomena, reducing the effective embedded length and consequently reducing the external stress that the fibre can stand, as shown in Figure 6.

Moreover, the absence of bending stiffness in carbon fibres, made up of several filaments that are aligned with one another and glued together by means of polymers, rules out the possibility of increases in pull-out load due to additional bending of the fibres which would compensate for the reduction connected with the local failure of the interface. This effect, described above, is particularly marked in the case of high strength matrices, in which the adding of silica fume creates a stronger interface with respect to the case of normal strength concrete, also increasing the tensile strength of the composite.

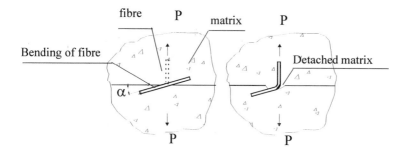

Figure 6 Possible mechanism of interface failure

Characteristic pull-out curve values for single carbon fibres

Table 3 gives the most significant values in the pull-out tests on carbon fibres from matrices with medium and high resistance, as a function of the inclination α and length L_f of the fibre. In particular, it shows the maximum load values P_{max} and the corresponding displacements δ_{max}. The same table also shows the values of the average tangential stress τ_{med} calculated as the ratio between the maximum pull-out load P_{max} and the lateral surface of the embedded fibre ($\pi \, \phi \, L_f/2$). These data obviously do not provide information about the effective distribution of tangential stresses in the fibre-matrix interface, which is far from being uniform [5]. However, the values recorded are very indicative in terms of order of magnitude. Lastly, Table 3 also gives the values of the overall energy expended by the external load to pull out the fibres, measured for the following characteristic displacement values: displacement corresponding to the value of the peak load (energy value indicated with E_0); the 3 mm displacement beyond which in all cases softening behaviour is expected (energy value indicated with E_{3mm}); displacement corresponding to a null load value (energy value indicated with E_t).

Table 3 Results of pull-out tests relating to carbon fibres

MATRICES	$\alpha°$	L_f (mm)	P_{max} (N)	δ_{max} (mm)	τ_{med} (N/mm^2)	E_0 (Nxmm)	E_{3mm} (Nxmm)	E_t (Nxmm)
NSC	0	20	31.00	0.28	1.26	8.28	78.76	143.44
NSC	0	30	50.00	0.44	1.36	19.62	100.76	219.79
HSC	0	20	50.92	0.22	2.07	10.77	110.30	173.45
HSC	22.5	20	42.12	0.40	1.72	14.86	66.49	141.98
HSC	45.0	20	35.81	0.38	1.46	10.16	58.13	114.78
HSC	67.5	20	19.10	0.40	0.78	7.30	33.00	44.00
HSC	0	30	67.00	0.41	1.82	21.60	169.69	393.64

NSC= normal strength concrete; HSC=high strength concrete

Comparison of Various Fibre Types

Figure 7 shows the curves of responses to pull-out of carbon, steel hooked and straight polyolefin fibres from high-resistance matrices.

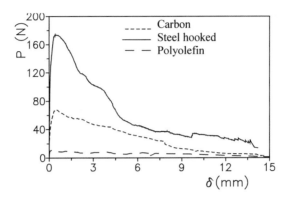

Figure 7 Effect of different types of fibre on pull-out from high-strength concrete

Examination of the graphs shows that the maximum pull-out load is that of steel hooked fibres, in which the longitudinal profile deformed by means of hooks increases resistance to pull-out with respect to the case of rectilinear fibres.

By contrast, one must consider the fact that in some cases, met with in the course of experimentation, for steel hooked fibres it happens that when the maximum load is reached there is failure of the transversal section of the fibre near the section attached to the hooked part. In this case, after this phase, progressive pull-out of the central stretch of the fibre continues. When failure does not occur, in most cases examined the fibre tends to straighten where there are hooks at the extremities and hence to pull out progressively, with brittle failure of the fibre-matrix interface, as can be evinced from the softening branches of the curves.

In the case of carbon fibres of the rectilinear type, though the maximum pull-out load is lower than that of hooked steel fibres, the overall behaviour proves to be more ductile. As soon as the maximum load is reached, the fibre begins progressively to pull out, but the slope of the softening branch is more attenuated than in the case of steel fibres. This result is probably due to the better surface state of multiple-filament fibres.

By contrast, the response of rectilinear polyolefin fibres certainly is less effective. The maximum pull-out load is 12 N, markedly lower than that of carbon or steel fibres, though the fibres exhibit ductile behaviour in the softening branch of the curve and progressive pull-out.

In Table 4, using the same symbols as used in Table 2, the most meaningful results of pull-out tests on the different types of fibres pulled out from high strength matrices are given.

Table 4 Results of pull-out tests on various types of fibres aligned with the external load (α=0) from high-resistance matrices

TYPE OF FIBRES	SHAPE	L_f (mm)	P_{max} (N)	δ_{max} (mm)	τ_{med} (N/mm^2)	E_0 (Nxmm)	E_{3mm} (Nxmm)	E_t (Nxmm)
Poyolefin		25	12.00	0.14	0.38	1.60	33.06	94.81
Steel		30	173.00	0.56	4.59	83.46	415.00	885.00
Carbon		30	67.00	0.41	1.82	21.60	169.69	393.64

CONCLUSIONS

A simple testing method to determine pull-out response of single fibres from cementitious matrices is presented. A comparison is made between pull-out of carbon, steel hooked and polyolefin fibres from normal and high strength concrete. The influence of embedded length, inclination angle of fibres with respect to external load, and type of matrices were investigated. The results obtained show the higher performances of steel hooked fibres in terms of strength with respect to carbon and polyolefin fibres. They also show the high level of energy dissipation for all the fibres investigated and especially the ductile behaviour of carbon fibres pulled out from high strength concrete. For this type of fibres this convenient response has to be considered in addition to other evident advantages like non-corrodibility and stability of the pysical properties.

REFERENCES

1. BENTUR, A, AND MINDESS, S. Fibre reinforced cementitious composites, Elsevier Applied Science, London & New york, 1990.

2. BRANDT, A, M. Cement-based composites: materials, mechanics properties and performance. E & FN Spon, London, Glasgow, Weinheim, New York, Melbourne, Madras, 1995.

3. BANTHIA, N, AND TROTTIER, J, F. Concrete reinforced with deformed steel fibers, part I: bond-slip mechanisms. ACI Material Journal,Vol 91, No 5, 1994, pp 346-356.

4. NAAMAN, A, E, NAMURM, J, M, A, AND NAJM, H, S. Fiber pullout and bond slip.II: experimental validation. ASCE Journal of Structural Division, Vol 117, No.9, 1991, pp 2791-2800.

5. CAMPIONE, G, MINDESS, S, AND ZINGONE, G. Failure mode in compression of fibre reinforced concrete cylinders with spiral steel reinforcement. Proceedings of BMC5 International Symposium, Warsaw, Polonia, 1997, pp 123-132.

IDENTIFICATION OF THE TYPE OF CEMENT IN HARDENED CONCRETE AND MORTARS

A Mueller

F Splittgerber

Bauhaus-University

Germany

ABSTRACT. Using the method shown in this paper it is possible to determine whether a Portland cement or a sulphate resistant Portland is used in a present concrete by dehydration of the separated paste at 1000 °C and a semi-quantitative evaluation of the XRD intensities of Tricalciumaluminate and Tetracalciumaluminateferrite. An identification of blast furnace slag cements should be possible by evaluation of Gehlenite but the influence of hydration time is currently being examined. Other types of cements can be identified by typical component for example Biotite and Nepheline as typical minerals of the Trass in a Trass cement.

Keywords: Type of cement, Hydration, Dehydration, Enrichment, Identification, Ordinary portland cement, Sulphate resistant portland cement, Blast furnace slag cements, Aggregates.

Prof Dr Ing habil Anette Mueller is Leader of the Chair of Processing of Building Materials and Reuse at the Bauhaus-University in Weimar, Germany. Her special interests are the recycling of construction debris for new building materials, the influence of processing on bulk materials and the chemistry of cements.

Dipl Ing Frank Splittgerber is Employee of the Chair of Processing of Building Materials and Reuse at the Bauhaus-University in Weimar, Germany. His special interests are the recycling of concrete and the chemistry of cements. He works on his dissertation titled "Identification of Cements Used in Hardened Concrete and Mortars".

INTRODUCTION

The determination of the type of cement used in a hardened concrete or mortar is important because damages appearing in buildings are frequently caused by an insufficient durability of the cement paste, that is, the use of unsuitable cements. Repairing these damages is very expensive and often leads to lawsuits between architects, building owners and users and builders and contractors. Considering this, it seems apparent that an unequivocal post-construction identification of the type of cement present in a concrete or mortar would be desirable.

LITERATURE

In the literature can be found descriptions of methods to determine the type of cement in practice, but none of them are very reliable, and sufficient only to distinguish cements relative to one item, e.g., sulphate resistance. One report [1] concerns determining the presence of blast furnace slag in a cement by the colour change of lead-acetate paper after treatment with H_2SO_4. Another report [2] contains a description of a procedure to classify various Portland cements by their source of generation, using trace elements. But, this works only with a knowledge of the raw materials being used by the cement factories. Another method [3] only allows distinguishing between an ordinary Portland cement and a high alumina cement. British Standard 1881 [4] specifies a procedure to determine the type of cement used in a hardened concrete by a combination of chemical and microscopical examinations. This method can be used to discern whether a Portland cement used in a concrete is sulphate resistant or not, but the method's reliability is quite low.

EXPERIMENTAL DETAILS

This work is directed at the question of whether or not the phases generated by the hydration of cements can be changed back into the primary cement phases. In the past there have only been statements about the theoretical aspects of the reversibility of cement hydration, specifically the hydration and dehydration of β-C_2S [5, 6]. The experimental results of the investigation are presented, and a possible method for identification of the type of cements is discussed.

The first step of the investigations involved the heat treatment, at various different temperatures, of powders made from hardened mortar specimens. The phases created thereby were analysed by x-ray diffraction (XRD). It was noted that at treatment temperatures greater than 1100 °C the samples frequently melted. This is not acceptable for XRD analysis because a recrystallization then is needed. For that reason several different temperature regimes were tried starting at 600 °C and increasing in steps of 100 K until 1100 °C. It was further observed that the XRD patterns obtained for samples that were heat treated at 1050 °C were most similar to those obtained from the original cement. Therefore this temperature was used for subsequent investigations.

For the second step, the enrichment of cement paste in the fine fractions was examined. In this paper we report only on the effect of enrichment achieved by conventional preparation methods. Specimen prisms 40 mm x 40 mm x 160 mm prepared from mortar containing several types of cement were used for these experiments. These prisms were crushed by a compressive strength testing machine and the coarse aggregate particles were removed by a 63 μm sieve. Despite this procedure the samples' content of insoluble Silica, caused by the aggregates, could not be decreased below 40%. More effective methods for enrichment of cement stone are being developed.

In the third step of the procedure, the prepared heat treated samples were examined by XRD. The results of XRD analyses of the samples treated at 1050 °C show the presence of cement phases or phases similar to cement phases. Larnite, Calcium Aluminate and Brownmillerite were identified after dehydration. In the cases of Calcium Aluminate and Brownmillerite an extraction with Salicylic Acid was needed to make them visible in XRD. Also, typical phases like Gehlenite and Merwinite, representing blast furnace slag, as well as Nepheline and Biotite, representing Trass, were found. The amounts of all the components were estimated in a semi-quantitative way by measuring and comparing the intensities of the XRD peaks. Thus, it was possible to distinguish between an ordinary and a sulphate resistant Portland cement. In some cases this estimation was not sufficient. For that reason other methods, e.g., the determination of free CaO, were employed.

XRD-Examinations

The Samples

Specimens from 17 different cements were examined. Most of them are Portland cements and blast furnace slag cements. An overview of these specimens is given in Table 1.

Table 1   Overview of the types of cements used for the examinations

NO.	TYPE OF CEMENT	ORIGIN
1	CEM I 32,5 R	Factory A
2	CEM I 32,5 R	Factory B
3	CEM I 32,5 R (sulphate resistant)	Factory B
4	CEM I 42,5 R	Factory C
5	CEM I 42,5 R (sulphate resistant)	Factory C
6	CEM I 42,5 R (sulphate resistant)	Factory A
7	CEM I 52,5 R	Factory C
8	CEM II/S 32,5	Factory D
9	CEM II/S 42,5	Factory D
10	CEM III 32,5	Factory C
11	CEM III 32,5	Factory D
12	CEM III 32,5 (sulphate resistant)	Factory E
13	CEM III 32,5 (low alkali content)	Factory E
14	CEM III 42,5	Factory D
15	CEM III 42,5	Factory E
16	CEM II/P 32,5	Unknown
17	CEM I 32,5 R + Pulverized Fly Ash	Factory B + F

RESULTS AND DISCUSSION

Specimens without Heat Treatment

In these samples Portlandite, Calcite and Ettringite could be proved as having crystalline phases. The only important one of these for identification purposes is Portlandite. The XRD intensities shown in Figure 1 present the clear differences in results found between the types of cements described in Table 1. It is observed that the largest amounts of Portlandite are contained in the samples made from Portland cements. This is due to higher contents of Tricalciumsilicate in Portland cement clinkers. Sample Numbers 12 and 13 show very low Portlandite contents because these two cements contain the largest amounts of blast furnace slag.

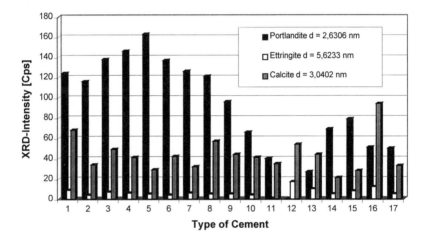

Figure 1   XRD intensities for Portlandite, Ettringite and Calcite determined at the non heat treated samples

In some samples, non-hydrated clinker phases like Larnite and Brownmillerite also were found. This probably occurred, despite the enrichment of the cement paste, because the remaining quartz from the aggregates was dominating the XRD patterns.

Specimens Heat Treated at 900 °C

Figure 2 shows the Dicalciumsilicate (C_2S) intensities determined before and after the dehydration by heat treatment at 900 °C. As can be seen C_2S can easily be identified in all samples. The difference in the XRD intensities of the Dicalciumsilicate before an after the heat treatment makes it probable that C_2S was created during the dehydration of C-S-H by the insertion of foreign oxides. The XRD peak of the Calciumsilicates is so very dominant in the XRD patterns of the heat treated samples, that the sample components Gehlenite, Anhydrite, Aluminate and Ferrite can only be found after the sample has been treated by an extraction of C_2S with Salicylic Acid.

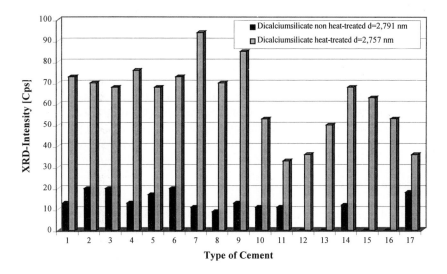

Figure 2   XRD intensities for C₂S determined at the non heat treated and at the
heat treated samples

In Figure 3 it can be observed, that the cements have differences in the XRD intensity of
Gehlenite. This substance, Gehlenite, is a main component of blast furnace slag and cannot be
found in Portland cements. Comparing the XRD results obtained with those cements containing blast furnace slag, it can be observed that the higher the content of blast furnace slag
in the cement, the higher the Gehlenite intensity in the XRD patterns.

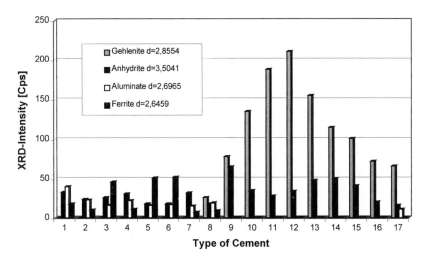

Figure 3   XRD intensities for Gehlenite, Anhydrite, Aluminate and Ferrite determined for the
heat treated samples after an extraction with Salicylic Acid

Where does all this Gehlenite come from? Is it a non-hydrated part of the blast furnace slag, which became visible by a recrystallization during heat treatment? Or, has it been generated by the dehydration of C-S-H? To answer this question, it is necessary to work with samples which are fully hydrated. However, for blast furnace slag cements this means a hydration time of at least 1 year. These tests are in progress.

In Figure 3 the differences in the Aluminate and Aluminate-Ferrite contents of the Portland cements are clearly visible too. Only the sulphate resistant ones (cements 3, 5 and 6) show higher intensities of Aluminate-Ferrite than of Aluminate.

Free Calcium Oxide Content

The content of free Calcium Oxide (CaO) of the heat treated samples was determined in order to evaluate the content of Tricalciumsilicate (C_3S) in the cements from which the samples were created. At a treating temperature of 900 °C, Tricalciumsilicate should not be generated. As C_2S was found after the heating process, the assumption was made that the difference in the content of CaO between C_3S and C_2S is present as free CaO, so that cements with higher C_3S contents should contain more free CaO after the heat treatment.

Evaluation of figure 4 shows that there is a difference between Portland cements (samples 1 to 5) and blast furnace slag cements (samples 8 to 15) but calculation of the content of C_3S in the cement from which the samples were created (the former cement) was not possible, because of the problem is that the standard EN 197 says that only 95 % of the cement have to be clinker, and if there is added e.g. limestone it will influence the results for free CaO.

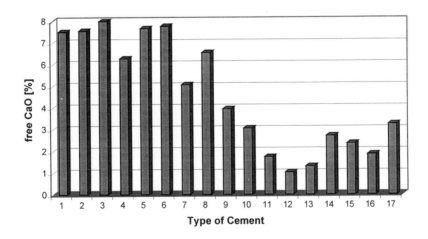

Figure 4   Contents of free Calcium Oxide (CaO) determined for the heat treated samples

SUMMARY

The described method for identification of the types of cements used in hardened concrete and mortars bases on mineralogical examinations of the heat treated cement pastes.

Using the XRD intensities for Aluminate and Aluminate-Ferrite it should be possible easily to distinguish Portland cements form sulphate resistant ones. Using the content of free lime an information about the amount of Tricalciumsilicate can be evaluated.

Concerning the blast furnace slag cements some more knowledge is needed. The success of the method primary depends on the determination of the amount the blast furnace slag. Further examinations with fully hydrated specimens will show whether the intensity of Gehlenite is suitable for that purpose.

Other types of cements could be identified by the presence of minerals which are typical for a certain type of cement, e.g. Nepheline and Biotite for the Trass in a Trass cement.

Additional works will deal with a determination of the compressive strength class of the cement.

We also continuing work on the development of procedures for a better enrichment of cement paste. Some progress was achieved in the last months. For example we have seen that it was possible to decrease the content of aggregate in the sample to less than 35 % by methods of conventional crushing and separation. Another method for separation being tested presently promises a further improvement.

Besides the work with real concrete and mortar there are ongoing studies with hydrated cement clinker phases and mixtures of them on the one hand and pure cement paste on the other hand to determine the general behaviour of cement paste during dehydration and the influence of the aggregates.

In the past examinations the type of cements containing significant amounts of Calcium Carbonate as well as concrete and mortars containing aggregates from Calcium Carbonate were excluded due to the fact that some methods used for all other types of cements, e.g. the determination of the insoluble residue representing the aggregates, can not be used at the presence of limestone.

REFERENCES

1. IKEN, H W. Handbuch der Betonpüfung.Düsseldorf: Beton-Verlag, 1994 S. 276-277.

2. GOGUEL, R L. St.John,D.A.Chemical identification of Portland cements in New Zealand concrete No. 1 & 2.Cem. Concr. Res. Vol. 23 S. 59-68 & 283-293.

3. BLANCO, M T, PUERTAS, F, VASQUEZ, T, DE LA FUENTE, A. The most suitable techniques and methods to identify high alumina cement and based Portland cement in concrete. Materiales de constuction. Madrid 42 (1992) S. 51-64.

4. BRITISH STANDARD INSTITUTION BS 1881. Testing Concrete. Part 124 November 1988. Methods for analysis of hardened concrete.

5. ISHIDA, H, SASAKI, K, OKADA, Y, MITSUDA, T. Hydration of β-C_2S prepared at 600 °C from Hillebrandite: C-S-H with Ca/Si = 1.9-2.0. 9th International Congress on the Chemistry of Cement New Delhi 1992 Vol. IV S. 76-82.

6. ISHIDA, H, MABUCHI, K, SASAKI, K, MITSUDA, T. Low-temperature synthesis of β-C_2S from Hillebrandite. J. Am. Ceram. Soc. 75 |9| 2427-2432 (1992).

MECHANISMS OF SULFATE ATTACK IN PLAIN AND BLENDED CEMENTS: A REVIEW

O S B Al-Amoudi

King Fahd University of Petroleum and Minerals

Saudi Arabia

ABSTRACT. The recent modifications in the physicochemical characteristics of Portland cements and the extensive use of mineral admixtures have introduced significant changes in the chemical and mineralogical composition in the present day cements that need to be addressed by the construction industry. Moreover, the effect of cation associated with the sulfate ions on these changes is inconclusive and highly debated in the concrete literature. In addition, the exposure of many reinforced concrete structures to sulfate-bearing environments has focussed attention on the role of sulfate ions in the corrosion of reinforcing steel. In view of the above-cited controversies, this paper presents a state-of-the-art review on sulfate attack in plain and blended cements from a civil engineering perspective. In particular, the paper addresses the following subjects: (i) mechanisms of sulfate attack in plain and blended cements; (ii) factors affecting sulfate attack; (iii) modes of deterioration; (iv) effect of sulfate cation type on the strength and expansion of plain and blended cements; (v) effect of sulfate ions on the weight loss and reinforcement corrosion of reinforced concrete specimens; (vi) role of chloride ions in sulfate attack in plain and blended cements; and (vii) protection of the concrete structures exposed to sulfate-bearing environments.

Keywords: Sulfate attack, Cements, Blended cements, Mortar, Concrete, Sodium sulfate, Magnesium sulfate, Expansion, Strength reduction, Weight loss.

Dr Omar Saeed Baghabra Al-Amoudi is an Associate Professor in the Department of Civil Engineering at King Fahd University of Petroleum and Minerals, Dhahran, Saudi Arabia. He obtained his Ph.D. degree from the same University in 1992. His current research interests are in the field of hot-weather concreting, durability of plain and blended cements in aggressive environments and protection of reinforced concrete using epoxy-coated rebars and corrosion inhibitors, non-destructive testing and stabilization of soils. He is a founding member of the ACI-Saudi Arabia Chapter and presently serving in the Board of Directors. Dr. Al-Amoudi has published many papers in reputed journals and conferences.

INTRODUCTION

Degradation of structural concrete components in sulfate-bearing environments has recently been concerned with the attack of sulfate ions and their associated cations on the products of hydration of portland cement, particularly the portlandite and calcium silicate hydrate [1-5]. In fact, the deterioration attributable to sulfate attack probably received attention much earlier compared to all other forms of deterioration such as, for example, corrosion of reinforcing steel or alkali-aggregate reactions [6]. A literature search indicates that Vicat, as early as 1818, reported a chemical attack on concrete due to the presence of sulfate ions in seawater, while Candlot established the formation of an expansive hydration product by the interaction of aqueous solutions of calcium aluminates and calcium sulfate in 1890 [7]. Michaelis, in 1892, attributed the disruption of concrete, when attacked by sulfate waters, to the reaction between C_3A in portland cement and sulfate ions to form ettringite.

Till today, research related to the effect of sulfate ions on the performance of concrete forms the second subject in the literature of concrete durability. Such a long and continuous research is motivated by the following factors: (i) modifications in the physicochemical characteristic of portland cements during the last four decades; (ii) the significant use of blending materials with portland cement to improve the performance of concrete when exposed to harsh environments; and (iii) the existence of some controversial points on the role of some parameters (i.e. cement content, C_3A, C_4AF) in sulfate attack [6]. Therefore, there is an urgent need to clarify the mechanisms of sulfate attack to the civil and constructional engineers, particularly those who specify blended cements in their reinforced concrete structures.

MECHANISMS OF SULFATE ATTACK ON PORTLAND CEMENTS

Research studies on sulfate attack are invariably conducted by exposing cement paste, mortar or concrete specimens to primarily sodium, magnesium and calcium sulfate solutions. Due to the limited solubility of calcium sulfate in water at normal temperatures (approximately 1400 mg/l SO_4^{2-}), higher concentrations are, therefore, due to the presence of magnesium sulfate $(M\overline{S})$ or sodium sulfate $(N\overline{S})$ [6]. When both or either of these salts exist in soils, seawater, sewage tanks, etc., there is an exigent need to understand the mechanisms of sulfate attack in order to protect the structural utilities when exposed to these hostile environments. In summary, the mechanisms of attack by $N\overline{S}$ and $M\overline{S}$ on hardened cement pastes are generally represented by the following reactions[*] [1,2,5-8]:

$$CH + N\overline{S} + 2H \rightarrow C\overline{S}H_2 + NH \qquad (1)$$

$$C_4AH_{13} + 3C\overline{S}H_2 + 14H \rightarrow C_6A\overline{S}_3H_{32} + CH \qquad (2)$$

$$C_4A\overline{S}H_{12} + 2C\overline{S}H_2 + 16H \rightarrow C_6A\overline{S}_3H_{32} \qquad (3)$$

$$C_3A + 3C\overline{S}H_2 + 26H \rightarrow C_6A\overline{S}_3H_{32} \qquad (4)$$

---

[*] $C = CaO$, $N = Na_2O$, $M = MgO$, $S = SiO_2$, $\overline{S} = SO_3$, $H = H_2O$

$$CH + M\overline{S} + 2H \rightarrow C\overline{S}H_2 + MH \tag{5}$$

$$C_xS_yH_z + xM\overline{S} + (3x + 0.5y - z)H \rightarrow xC\overline{S}H_2 + xMH + 0.5yS_2H \tag{6}$$

$$4MH + SH_n \rightarrow M_4SH_{8.5} + (n - 4.5)H \tag{7}$$

Mechanisms of $N\overline{S}$ Attack

This attack is initiated by the reaction of $N\overline{S}$ with the portlandite (CH) produced by the hydration of cement. The sodium hydroxide (NH) produced by this reaction (Equation 1 above) raises the pH value of the cement paste to 13.5 (i.e. compared with 12.4 for saturated CH solutions), thereby stabilizing both C-S-H and ettringite ($C_6A\overline{S}_3H_{32}$). The gypsum ($C\overline{S}H_2$) produced by Equation (1) will react with some of the hydration products, namely calcium aluminate hydrate (C_4AH_{13}), monosulfate ($C_4A\overline{S}H_{12}$) and/or unhydrated (i.e. surplus) tricalcium aluminate (C_3A), to produce "secondary" ettringite, as summarized in Equations (2) to (4). Since ettringite is expansive in nature (i.e. it has a low density of 1.73 g/cm^3 compared to an average of 2.50 g/cm^3 for the other products of hydration), expansion and cracking of concrete are the manifestation of $N\overline{S}$ attack [9].

In summary, $N\overline{S}$ attack is based on the generation of secondary ettringite as per Equations (2) to (4). The reactants in these equations are truly aluminate based (i.e. monosulfate, hydrated aluminate and/or unhydrous C_3A). Thus, it is rational to specify a C_3A content less than 5% to produce $N\overline{S}$-resistant cements. If the C_3A content is between 5 and 8%, the $N\overline{S}$ attack is possible. If the C_3A is more than 8%, the attack will be probable. There are two more factors that influence this type of attack; the first being the content of C_4AF (tetracalcium alumino-ferrite). Although this mineral behaves less deleteriously than C_3A in terms of $N\overline{S}$ attack, it contributes to the development of a phase similar to, but less expansive than, ettringite [9]. Therefore, ASTM C 150 limits the total content of C_4AF plus twice the C_3A content to 20%. Secondly, the NH produced by Equation (1) has a stabilizing effect on ettringite thereby making it more expansive and detrimental. Further, the gypsum produced by Equation (1) has more volume than that of the reactants [8]. Accordingly, if a $N\overline{S}$-resistant cement is to be produced, the portlandite (CH) content should be reduced (to retard Eqn. 1) in addition to the ASTM C 150 limitations on C_3A and C_4AF [10]. CH can be reduced by reducing the C_3S/C_2S ratio or using blended cements [3].

Mechanisms of $M\overline{S}$ Attack

This attack starts by Reaction 5 above. However, unlike NH, the magnesium hydroxide (MH) produced in this reaction is insoluble (i.e. its solubility is 0.01 g/l compared to 1.37 g/l for CH) and its saturated solution has a pH value of 10.5 (compared to a pH of 12.4 and 13.5 for CH and NH, respectively) thereby destabilizing both ettringite and C-S-H. The consequences of this low pH are [8-10]: (i) secondary ettringite will not form; (ii) $M\overline{S}$ will readily react with C-S-H, as per Equation (6) above, to produce gypsum, brucite (MH) and silica gel (S_2H). This gel is less cementitious than the original cementing C-S-H gel; (iii) C-S-H tends to liberate lime to raise the pH. The liberated lime reacts further with $M\overline{S}$ (as per Reaction 5 above) and, therefore, produces more MH; accordingly (iv) the concentration of gypsum and brucite in the hardened paste matrix will increase while the C-S-H progressively loses its lime and becomes less cementitious; and (v) the deleterious action of MH is ascribable to its reaction with the hydrosilicates (S_2H), as shown in Equation (7), thereby producing magnesium silicate hydrate (M-S-H), which is non-cementitious.

Therefore, the damaging attribute of $M\overline{S}$ stems from the fact that Reactions 6 and 7 go on to completion thereby converting the C-S-H phase to a fibrous, amorphous material (M-S-H) with no binding properties [8]. This attack is, therefore, characterized by softening and deterioration of the surficial layers of the hardened cement paste and the profuse formation of gypsum and brucite (Equations 5 and 6). However, the brucite content will ultimately decrease due to its conversion to M-S-H (Equation 7).

FACTORS AFFECTING SULFATE ATTACK

The above-cited mechanisms indicate that all cements (including Type V one) are vulnerable to $N\overline{S}$ and/or $M\overline{S}$ attack. However, the degree of attack depends on the following factors: (i) cement type (i.e. C_3A, C_3S/C_2S ratio and C_4AF); (ii) sulfate type and concentration; (iii) quality of concrete (i.e. permeability); (iv) form of construction (i.e. thin vs. thick section and exposure to wet-dry cycles); (v) level of water table and its seasonal variation (i.e. capillary action and salt crystallization); and (vi) possibility of replenishment.

MODES OF SULFATE ATTACK

A review of the literature [7] indicates that three modes of deterioration are usually manifested with sulfate attack. The first mode is akin to eating away of the hydrated cement paste and progressively reducing it to a cohesionless granular mass leaving the aggregate exposed and leading to loss of mass and reduction in strength. This mode is attributed mainly to the formation of gypsum, and is known as the acidic type of $M\overline{S}$ attack (see Figure 1). The second mode, which is normally characterized by expansion and cracking, takes place when the reactive hydrated aluminate phases, present in sufficient quantities, are attacked by $N\overline{S}$ ions, thereby forming tricalciumsulfo-aluminate hydrate, also called ettringite or Candlot's salt. This expansive type of reaction is ascribable to the formation of a colloidal form of ettringite in the presence of high pH in the pore solution. The third mode of sulfate attack is the onion-peeling type, which is characterized by scaling or shelling of the surface in successive layers in the form of "onion-skinning" delamination [11]. This mode is the least reported in the literature and was observed in the case of paste specimens made with plain and fly ash blended cements exposed to mixed-sulfate environments [7].

SULFATE ATTACK ON BLENDED CEMENTS

In the context of this paper, blended cements are those portland cements to which mineralogical (i.e. pozzolanic) admixtures are added either as a replacement of the parent cement or as an addition to it [12].

The interaction of pozzolanic materials with the products of hydration of portland cements is known as the pozzolanic reaction, which can be simplified by the following equation:

$$
\begin{array}{ccccc}
& & & H_2O & \\
\text{Silica} & + & \text{Portlandite} & \xrightarrow{\hspace{1cm}} & \text{Secondary C-S-H} \quad (8)\\
\text{(from pozzolan)} & & \text{(from cement)} & \text{Normal Temperature} &
\end{array}
$$

The "secondary" calcium-silicate hydrate is similar in essence to the "primary" C-S-H produced by the hydration of portland cement itself though the former is less dense [10]. This pozzolanic reaction has the following beneficial impacts on sulfate attack: (i) the consumption of portlandite in Eqn. (8) reduces the formation of gypsum in Equations (1) and (5). Therefore, both $N\overline{S}$ and $M\overline{S}$ attacks are alleviated; (ii) the replacement of part of the cement by a pozzolanic material entails a reduction in the C_3A content (i.e. dilution effect). Hence, all the aluminate-bearing phases will accordingly be reduced.   Therefore, the formation of secondary ettringite [as per Equations (2) to (4)] will be mitigated; (iii) even if it is formed, ettringite becomes expansive only at high pH values.   Since blended cements consume most of the portlandite produced by cement hydration and reduce the pH, the ettringite becomes less expansive; (iv) the formation of secondary C-S-H produces a film or a coating on the alumina-rich and other reactive phases thereby hindering the formation of secondary ettringite; and (v) the formation of secondary C-S-H also results in the densification of the hardened cement paste since it is deposited in the pores thereby making blended cements impermeable and, therefore, sulfate ions cannot easily penetrate through the concrete matrix.

The above advantages are truly effective in $N\overline{S}$ environments only where the sulfate attack follows Equations (1) to (4) because the formation of secondary ettringite (as well as gypsum in Eqn. 1) will be mitigated. However, the situation is totally different when blended cements are exposed to $M\overline{S}$ environments whereby the attack follows Reactions 5, 6 and 7. It is true that Reaction 5 will be mitigated by the usage of pozzolanic materials (in a way similar to Reaction 1 in $N\overline{S}$ attack) due to the absence or reduction of portlandite by the pozzolanic reaction (Equation 8). However, instead of reacting with portlandite in a way similar to the attack on plain cements, $M\overline{S}$ will directly and energetically attack the cementitious C-S-H and convert it to the non-cementitious M-S-H. Such an attack is very instrumental to the enhanced deterioration of blended cements.

In summary, the above mechanisms indicate that blended cements are more resistant to $N\overline{S}$ attack than plain portland cements. However, these blended cements are more susceptible to attack by $M\overline{S}$. Although portlandite (CH) is "bad" in $N\overline{S}$ environments, it is very beneficial in $M\overline{S}$ because CH constitutes the first line of defense against $M\overline{S}$ attack and, therefore, it serves to protect the C-S-H. However, in the long term, all cements, whether blended or plain, may suffer equally from $M\overline{S}$ attack. Therefore, protection of the substructures exposed to $M\overline{S}$ environments should be augmented.

MECHANICAL PROPERTIES OF MORTAR SPECIMENS

To assess the mode and degree of degradation due to exposure to $N\overline{S}$ and $M\overline{S}$ environments, the strength reduction and expansion of mortar specimens made with nine plain and blended cements were monitored for a period of 12 months [10]. The strength reduction was determined by comparing the compressive strength of small (25 mm) cubes in the sulfate environments with the strength of similar specimens cured in water. The expansion measurements were conducted on prismatic (25×25×285 mm) mortar specimens. The water to binder or cementitious materials ratio (w/b) was maintained at 0.50 except for Mix # 7, as shown in Table 1, while the sand to binder ratio was maintained at 2.75. The sulfate (SO_4^{2-}) concentration was maintained at 2.1%.

Table 1 Strength reduction and expansion in $N\overline{S}$ and $M\overline{S}$ environments [10]

MIX NO	CEMENT TYPE	BLENDING MATERIAL	W/B**	STRENGTH REDUCTION, after 1 year (%)			EXPANSION, after 1 year (%)	
				$N\overline{S}$	$M\overline{S}$	Difference	$N\overline{S}$	$M\overline{S}$
1	I	-	0.50	39	62	23	0.104	0.046*
2	V	-	0.50	34	53	19	0.113	0.031
3	I	FA (20 %)	0.50	22	65	43	0.111	0.056*
4	I	SF (20 %)	0.50	9	72	63	0.082	0.096
5	I	SF (20 %)	0.50	−1	89	90	0.061	0.090*
6	I	BFS (70 %)	0.50	18	75	57	0.099	0.067*
7	I	-	0.35	26	81	55	0.095	0.088
8	V	FA (20 %)	0.50	17	66	49	0.089	0.086
9	V	SF (10 %)	0.50	7	71	64	0.073	0.082

*Expansion measurements taken earlier than 360 days.
**Water-to-binder ratio

Strength Reduction

The strength reduction data in Table 1 indicates that all blended cements exhibited superior performance in $N\overline{S}$ environment as compared with plain cements. However, silica fume (SF) cement displayed distinctly the best performance whereby the reduction in strength ranged from −1% to 9% (for 20 and 10% SF additions, respectively). The negative strength reduction proves the very dense microstructure of the 20% SF cement whereby the strength of the specimens exposed to $N\overline{S}$ solution was higher than that of the specimens exposed to water due to the densifying effect of the pozzolanic reaction products in the existing dense microstructure. Further, comparing the performance of Mix No. 4 and 9 (both of them have 10% SF) indicates that SF cements do not significantly depend on the type of parent cement (i.e. whether Type I or Type V).

Blast furnace slag (BFS) cement (Mix No. 6) showed the second best performance while fly ash (FA) cement exhibited an intermediate performance between blended and plain cements. In fact, the performance of FA cement (Mix No. 3 and 8) was equivalent to the cement made with low w/b ratio (Mix No. 7).

On the contrary to $N\overline{S}$, the strength reduction was very high in all the cements exposed to $M\overline{S}$ solution. However, all blended cements exhibited greater reduction in strength than plain cements. To quantify the disparity in performance of the cements, the difference in strength reduction in $M\overline{S}$ and $N\overline{S}$ exposures was included in Table 1. SF-blended cements displayed uniquely extreme difference in performance in the two exposures (i.e. from the best in $N\overline{S}$ to the worst in $M\overline{S}$) followed by BFS, low w/b ratio, and FA cements. These results prove that blended cements, particularly those made with SF and BFS, are very deleteriously affected by $M\overline{S}$, primarily due to their low content of portlandite [5-8].

The inferior performance of the low w/b ratio specimens in $M\overline{S}$ environment was probably attributed to their dense microstructures thereby providing limited space for the expansive reaction products to occupy (see [6,10]). However, $N\overline{S}$ attack predominantly proceeds via penetration of the sulfate ions into the interconnected porosity and the strength reduction takes place mainly when the reaction products are distributed into the whole specimens and partly by the leaching of soluble reaction products [13]. Accordingly, such a process is hindered in the low w/b ratio cement due to its dense microstructure and better resistance to $N\overline{S}$ penetration. In $M\overline{S}$ exposure, however, the extensive gypsum precipitation (Equations 5 and 6) and the massive M-S-H formation (Eqn. 7) mostly occurred on the surfaces of the specimens leading to loss of material and reduction in strength; this damaging attribute tends to keep up with the penetrating front of $M\overline{S}$ [13]. This attack is accentuated in dense matrices such as those of low w/b ratio and blended cements.

Expansion

The expansion data in Table 1 clearly indicates that, except for SF cements which exhibited significant deterioration, all plain and blended cements exposed to $N\overline{S}$ solution displayed higher expansion than those in $M\overline{S}$. Furthermore, blended cements, particularly those made with SF, exhibited the best performance in $N\overline{S}$ as compared with plain cements. Therefore, the trend of the expansion results is almost exactly similar to the strength reduction of the specimens exposed to $N\overline{S}$. Surprisingly, the lower values of expansion of the specimens exposed to $M\overline{S}$ environment do not commensurate with the data on strength reduction. Further, the lower expansion of the specimens exposed to $M\overline{S}$, as compared with $N\overline{S}$, indicates that the deteriorating mechanisms therein do not produce expansive products, particularly in the interior portions (i.e. cores) of the specimens. Consequently, the degradation of mortar specimens is proven to be superficial and does not extend to the core of the specimens. This conclusion is also evidenced by the fact that some of the stainless steel studs on the sides of the expansion specimens were loosened by the extensive deterioration leading to termination of the test measurements before the 360 days [10].

If a threshold value for expansion is taken as 0.10% [10], the expansion data in Table 1 clearly indicates that none of the plain and blended cements exposed to $M\overline{S}$ environment reached failure, even after 360 days of exposure. Furthermore, in $N\overline{S}$ solution, only plain Type I and Type V cements as well as FA cement (with Type I) passed the threshold expansion value of 0.1% after 12 months. This group of cements is followed by BFS cement, low w/b ratio cement and FA cement (with Type V). The lowest expansion readings were attained by the SF cements. This classification of cements agrees very well with the strength reduction data in $N\overline{S}$, which clearly indicates the superiority of blended cements; specifically the SF ones followed by BFS cement, in resisting $N\overline{S}$ attack.

DURABILITY OF REINFORCED CONCRETE SPECIMENS

In order to assess the effect of sulfate attack on plain and blended cements, fifteen reinforced concrete mixtures were exposed to a mixed sulfate solution [6]. Details of the plain and blended cements, w/b ratio, etc. are presented in Table 2. The concentration of test solution was kept as 2.1% (SO_4^{2-}); similar to that used in Table 1. In this investigation, the weight loss of concrete and the degree of corrosion of reinforcing steel were assessed after 44 months. The threshold values of weight loss, corrosion potential and polarization resistance are 5%, −270 mV and 87 kΩcm^2, respectively [6]. The data in Table 2 indicates that all the concrete specimens made with blended cements have already passed the failure criterion of 5% weight loss. On the contrary, plain cement concretes made with a w/b ratio of 0.50 exhibited minor weight losses of less than 5%. Therefore, all plain cements (Type I, Type II and Type V) performed very well despite thefact that their C_3A content varied from 3.5% to 8.5%. However, when their w/b ratio decreased from 0.50 to 0.35, the weight loss accordingly increased (i.e. compare Mix No. 1 and 2 with 7 and 10).

Table 2 Long-term data on the durability of reinforced concrete mixtures [6]

MIX NO.	CEMENT TYPE	BLENDING MATERIAL (REPLACEMENT)	W/B**	CONCRETE WEIGHT LOSS (%)	CORROSION POTENTIAL, SCE (−mV)	POLARIZATION RESISTANCE (kΩcm^2)
1	I	None	0.50	0.86	395.1	56.7
2	V	None	0.50	0.93	337.7	76.6
3	I	Fly Ash (20%)	0.50	23.1	681.1	24.1
4	I	Silica Fume (10%)	0.50	8.99	252.1+	844+
5	I	BFS* (60%)	0.50	37.4	696.2	12.8
6	II	None	0.50	3.13	577.9	70.7
7	I	None	0.35	8.88	259.4+	696+
8	V	Fly Ash (20%)	0.50	11.6	670.7	8.92
9	V	Silica Fume (10%)	0.50	14.8	652.2	30.5
10	V	None	0.35	4.72	188.5+	386+
11	I	BFS* (60%)	0.35	44.0	669.4	6.81
12	I	Fly Ash (20%)	0.35	18.5	713.0	27.4
13	V	Fly Ash (20%)	0.35	11.0	668.8	44.2
14	I	Silica Fume (10%)	0.35	11.6	503.3	162+
15	V	Silica Fume (10%)	0.35	15.5	350.8	327+

*Blast furnace slag   +Passive corrosion state
**Water-to-binder ratio

Similarly, when the w/b ratio of all blended cement concrete specimens decreased, their weight loss increased (i.e. compare Mix No. 3, 4, 5, 8 and 9 with 12, 14, 11, 13 and 15, respectively). Therefore, it can be concluded that there is a detrimental effect associated with the reduction in w/b ratio of all plain and blended cement concretes. This finding is in parallel agreement with the results of strength reduction in plain cement mortar specimens in the $M\overline{S}$ environment presented previously (Table 1). Furthermore, the mode of deterioration in mixed-sulfate exposures is predominantly controlled by $M\overline{S}$ attack [7,8]; typically manifested by softening of the surficial layers and reduction in strength and loss of weight (i.e. the acidic mode of attack). Visual documentation of the fifteen reinforced concrete mixtures graphically confirms this conclusion (Figure 1).

As shown in Table 2, the data therein indicates that most of the steel bars were in an active state of corrosion because the corrosion potentials were more negative than −270 mV with respect to saturated calomel electrode and the polarization resistance was less than 87 kΩcm^2. Even those mixtures which exhibited negligible deterioration (i.e. Mix No. 1 and 2) could not protect the passivity of steel. For blended cement concretes, the corrosion was certainly attributable to the significant weight loss (i.e. more than 10%) whereby their concrete cover was tremendously diminished by $M\overline{S}$ attack. However, the SF cement concrete specimens (Mix No. 4) displayed distinguished performance. Despite its high weight loss (9%), the steel in SF cement concrete was still passive and the sulfate ions could not ingress through during the 44 months of exposure. Such excellent performance of SF concretes proves the fact that *the core of these materials was dense and intact despite the extensive surficial deterioration.* Similar conclusions can be said about the low w/b ratio

plain cement concretes (Mix No. 7 and 10). For BFS cement, the concrete specimens as well as their reinforcing steel exhibited extremely inferior performance - much worse than anticipated. Similar inferior performance of BFS cement specimens was observed in sulfate and sulfate-chloride exposures, as is discussed in the subsequent part. The reasons could be partly attributed to the high MgO and SO_3 contents (8.8% and 3.1%, respectively) of BFS. Details on the inferior performance of BFS cements are reported elsewhere [6,14].

ROLE OF CHLORIDES IN SULFATE ATTACK

The performance of nine plain and blended cement mortar specimens was evaluated by measuring the strength reduction (using 50 mm cubes) and expansion in pure sulfate environment (SO_4^{2-} = 2.1%) and in sulfate-chloride environment (SO_4^{2-} = 2.1% and Cl^- = 15.7%) [15]. The sulfate type and concentration were exactly similar to that used in Table 2. Table 3 summarizes the strength reduction and expansion after one year of exposure to these two solutions. The data therein clearly indicates that the concomitant presence of chlorides with the sulfate ions mitigated the strength reduction. However, the extenuation of sulfate attack by chloride addition was more pronounced in plain cements (Mix No. 1, 2, 5 and 6) as well as in fly ash cements (Mix No. 3 and 7). In contrast, the chloride-triggered mitigation was marginal in blended cements made with SF and BFS. Based on microstructural analyses using x-ray diffraction and scanning electron microscopy, Al-Amoudi et al. [16] reported that the significant mitigation of sulfate attack (i.e. termed chloride beneficiation) on plain cements was ascribable to the inhibition effect of chlorides on the expansive gypsum- and ettringite-oriented types of sulfate attack, which are predominantly operative in plain cements

(a) Mixes # 1 to 5

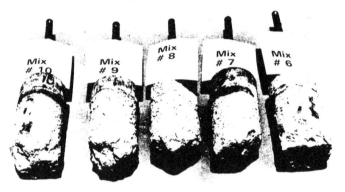

(b) Mixes # 6 to 10

(c) Mixes # 11 to 15

Figure 1 Visual documentation of the various reinforced concrete mixtures

(Reactions 1 to 5 above). Such inhibition was also manifested by the low values of expansion of all cements in the sulfate-chloride exposure (see Table 3). Contrarily, the beneficial effect of chlorides was marginal on blended cements because of the inability of chlorides to inhibit the Mg-oriented type of sulfate attack (Equations 6 and 7) [16].

Table 3 Strength reduction and expansion data in sulfate-chloride environments [15]

MIX NO.	CEMENT TYPE	BLENDING MATERIAL	W/B*	STRENGTH REDUCTION (%)		EXPANSION (%)	
				Sulfate solution	Sulfate-chloride solution	Sulfate solution	Sulfate-chloride solution
1	I	None	0.50	42	16	0.106	0.045
2	V	None	0.50	48	17	0.093	0.046
3	I	Fly ash (20%)	0.50	47	9	0.099	0.039
4	I	Silica fume (10%)	0.50	53	41	0.094	0.036
5	I	None	0.35	48	14	0.114	0.040
6	V	None	0.35	54	26	0.096	0.046
7	V	Fly ash (20%)	0.50	55	16	0.088	0.025
8	V	Silica fume (10%)	0.50	49	40	0.093	0.033
9	I	BFS** (70%)	0.50	55	45	0.113	0.059

*Water-to-binder ratio   ** Blast furnace slag

Despite their inferior performance in sulfate-chloride exposures (Table 3), blended cements made with SF and BFS have shown superior resistance to corrosion of reinforcing steel, much more than all the other plain concrete mixtures [17]. As corrosion of reinforcing steel is the predominant mode of deterioration in chloride-sulfate exposures (i.e. sabkha soils, coastal and offshore structures), it is rational to specify SF-blended cement in such aggressive media to retard or inhibit the corrosion of reinforcing steel. However, to mitigate $\overline{MS}$ attack on the concrete material, additional protective measures should be warranted.

PROTECTION OF CONCRETE

The data in Tables 1 to 3 indicates that one of the most important factors playing keynote role in the protection strategy is the existing type of sulfate attack (i.e. whether $\overline{NS}$ or $\overline{MS}$). The presence of concomitant chloride should also be thoroughly considered because of its influence on both the sulfate attack and corrosion of reinforcing steel.   However, the following protective measures can be considered very instrumental in the resistance against sulfate attack: (i) try to avoid the sulfate exposures, if possible, through the usage of geotextiles, water drainage, capillary cut-off, tanking the foundation, etc.; (ii) use sulfate-resistant (Type V) cement with a relatively low C_3S/C_2S ratio. If chlorides are conjointly present with the sulfates, use SF-blended cement.   If the magnesium concentration is comparatively high, use a water-resistant epoxy-based coating on the exterior surfaces that are in direct contact with the solution; (iii) use dense concrete (i.e. w/b ratio of 0.35 to 0.40) with a relatively high cement content (i.e. about 400 kg/m^3); and (iv) use thick structural members and keep some additional part of the cover as a sacrificial layer (i.e. about 10 mm).

CONCLUSIONS

This paper presented a comprehensive review on the effect of sulfate attack on plain and blended cements. Blended cements, particularly those made with SF, were highly resistant to $N\overline{S}$ attack due to an interplay of several factors; the most important being the reduction in the portlandite produced by cement hydration and the densification of the microstructure of hardened cement paste. These factors mitigate the production of expansive ettringite and gypsum. Low w/b ratio cements moderately alleviate $N\overline{S}$ attack by mitigating the penetration of SO_4^{2-} ions into the dense matrix of the specimens. $N\overline{S}$ attack is manifested by excessive expansion and marginal reduction in compressive strength.

Upon exposure to $M\overline{S}$, blended cements, particularly those made with SF and BFS, displayed inferior performance in terms of strength reduction and weight loss of concrete. Their inferior performance is ascribable to the consumption of portlandite by the pozzolanic reaction thereby exacerbating the attack on C-S-H, leading to softening, loss of concrete material and excessive reduction in strength. Ultimately, $M\overline{S}$ attack transforms the cementitious C-S-H into non-crystalline M-S-H with no cementing properties.

Despite its excessive deterioration in $M\overline{S}$ exposures, SF cement succeeded to protect the reinforcing steel more than all the other plain and blended cements. This superior performance indicates that the $M\overline{S}$ attack was only surficial and the core of SF cement concrete specimens remained dense and impermeable. Similar excellent performance of this cement has been observed in sulfate-chloride exposures. Therefore, SF cement has been recommended in these aggressive environments. However, if the concentration of Mg ions is high, additional protection (i.e. usage of epoxy-based coating) should be incorporated.

The concomitant presence of chlorides tends to mitigate $N\overline{S}$ attack due to the enhanced solubility of gypsum and ettringite thereby inhibiting their expansive characteristics. In $M\overline{S}$ exposures, chlorides alleviate the gypsum attack (similar to that in $N\overline{S}$ environments), however, they do not significantly affect the attack of $M\overline{S}$ on C-S-H and convert it to non-cementitious M-S-H.

ACKNOWLEDGEMENT

The support of the Department of Civil Engineering at KFUPM is appreciated. This paper has recently been presented at the Symposium on the Performance of Concrete Structures in the Arabian Gulf Environment, KFUPM, November 15-17, 1998.

REFERENCES

1.   MATHER, M.   Field and laboratory studies of the sulphate resistance of concrete. Performance of Concrete, Ed. E G Swenson, University of Toronto Press, 1968, pp 66-76.

2.   CALLEJA, J.   Durability.   Proc. 7th Int. Congress on Chemistry of Cement, Paris, 1980, Editions Septima, Sub-Theme VII-2, Vol.1, pp VII:2/1-VII:2/48.

3.   RASHEEDUZZAFAR. Influence of cement composition on concrete durability. ACI Materials Journal, 1992, Vol.89, No.6, pp 574-586.

4.   LAWRENCE, C D. Sulfate attack on concrete. Mag. Concr. Res., 1990, Vol.42, No.153, Dec., pp 249-264.

5.   COHEN, M D AND BENTUR, A. Durability of portland cement-silica fume pastes in magnesium sulfate and sodium sulfate solutions. ACI Materials Journal, 1988, Vol.85, No.3, May-June, pp 148-157.

6.   AL-AMOUDI, O S B. Performance of fifteen reinforced concretes in magnesium-sodium sulphate environments. Constr. and Bldg. Materials, 1995, Vol.9, No.1, pp 25-33.

7.   AL-AMOUDI, O S B. Sulfate attack and reinforcement corrosion in plain and blended cements exposed to sulfate environments. Building and Environment, 1998, Vol.33, No.1, pp 53-61.

8.   RASHEEDUZZAFAR, AL-AMOUDI, O S B, ABDULJAUWAD, S N AND MASLEHUDDIN, M. Magnesium-sodium sulfate attack in plain and blended cements. ASCE Journal of Materials in Civil Engineering, 1994, Vol.6, No.2, May, pp 201-222.

9.   LEA, F M. The Chemistry of Cement and Concrete, 3rd Edition, Edward Arnold, London, 1970.

10. AL-AMOUDI, O S B, MASLEHUDDIN, M AND SAADI, M M. Effect of magnesium sulfate and sodium sulfate on the durability performance of plain and blended cements. ACI Materials Journal, 1995, Vol.92, No.1, pp 15-24.

11. FIGG, J W. Chemical attack on hardened concrete, effect of sulphates and chlorides. Bull. Inst. of Corrosion Sci. and Tech., 1979, No.75, July, pp 12-23.

12. MEHTA, P K. Pozzolanic and cementitious by-products as mineral admixtures for concrete - A critical review. ACI SP-79, American Concrete Institute, Detroit, 1987, Vol.1, pp 1-46.

13. ISRAEL, D E, MACPHEE, D E AND LACHOWSKI, E E. Acid attack on pore-reduced cements. Journal of Materials Science, 1997, Vol.32, No.15, Aug., pp 4109-4116.

14. GOLLOP, R S AND TAYLOR, H F W. Microstructural and microanalytical studies of sulfate attack: IV. Reactions of a slag cement paste with sodium and magnesium sulfate solutions. Cem. and Concr. Res., 1996, Vol.26, No.7, pp 1013-1028.

15. AL-AMOUDI, O S B, MASLEHUDDIN, M AND ABDUL-AL, Y A B. Role of chloride ions on expansion and strength reduction in plain and blended cements in sulphate environments. Constr. and Bldg. Materials, 1995, Vol.9, No.1, Feb., pp 25-33.

16. AL-AMOUDI, O S B, RASHEEDUZZAFAR, MASLEHUDDIN, M AND ABDULJAUWAD, S N. Influence of chloride ions on sulphate deterioration in plain and blended cements. Mag. Concr. Res., 1994, Vol.46, No.167, June, pp 113-123.

17. AL-AMOUDI, O S B, RASHEEDUZZAFAR, MASLEHUDDIN, M AND ABDULJAUWAD, S N. Influence of sulfate ions on chloride-induced reinforcement corrosion in plain and blended cement concretes. Cement, Concrete, and Aggregates, 1994, Vol.16, No.1, pp 3-11.

PERFORMANCE OF BLENDED CEMENT CONCRETES IN A MARINE ENVIRONMENT

T H Wee   S F Wong   H B Lim

The National University of Singapore

P K Mah   K S Chia

Pan Malaysia Cement Works (Singapore) Pte Ltd

Singapore

ABSTRACT. This paper presents a study on the performance of Portland cement (PC) concretes with and without mineral admixtures in marine environment. Chloride diffusion and cube tests are conducted at 20, 30 and 40°C to investigate the chloride ingress and strength development in temperate, tropical and hot climates respectively. Mercury intrusion porosimetry is performed to study the microstructure of concrete. The rates of chloride ingress and strength development, as well as pore size distribution, are observed to be dependent on ambient temperature, water-binder ratio and use of mineral admixtures in concrete. The initiation period for chloride-induced corrosion of steel reinforcement in concrete, in relation to the enhancement of service life, is also determined to assess performance of marine structures.

Keywords: Portland cement (PC), Mineral admixtures, Marine, Climate, Chloride ingress, Strength development, Microstructure, Service life, Performance.

Dr T H Wee is an Associate Professor in the Department of Civil Engineering, The National University of Singapore. His research interests include durability of construction materials; microstructural and thermal properties of concrete.

Dr S F Wong is a Research Fellow in the Department of Civil Engineering, The National University of Singapore. Her research activities cover both experimental and analytical work on the resistance of concrete against chemical attacks (e.g. chloride ingress, sulphate attack and carbonation); microstructural and thermal properties; and use of admixtures in concrete.

Mr P K Mah is the General Manager of Pan Malaysia Cement Works (Singapore) Pte Ltd. He oversees the production and supply of high-quality cement and mineral admixtures to both local and overseas construction industries.

Mr K S Chia is the Works Manager of Pan Malaysia Cement Works (Singapore) Pte Ltd. He has vast experience in the selection of suitable cementitious materials for use in concrete structures

Mr H B Lim is a Principal Laboratory Technologist in the Department of Civil Engineering, The National University of Singapore. He has more than twenty years working experience in Concrete Technology.

.

INTRODUCTION

In recent years, the development of blended cement concretes is becoming increasingly important in the construction of reinforced concrete (RC) structures exposed to different marine environments and aggressive climatic conditions. The performance of concrete is not only characterised by its compressive strength, but also by its durability in relation to its resistance against aggressive chemical agents such as chloride ions (Cl⁻) [1], and its microstructure [2].

The use of mineral admixtures has been reported [2-4] to be one of the most efficient methods to improve the resistance of concrete to Cl⁻ ingress. Collepardi et al. [5] observed higher resistance to Cl⁻ ingress in pozzolanic cement as compared to Portland cement (PC), while research work on ground granulated blast-furnace slag (GGBS) [2,4] and silica fume (SF) [6,7] shows better resistance of concrete to Cl⁻ ingress with increasing replacement percentage of cement by GGBS or SF.

Studies by Byfors [4] reveal that a higher fineness of GGBS results in a larger specific surface area and thus a higher reactivity of GGBS. This leads to concretes with smaller pore volumes, higher rates of microstructural densification and cement hydration which in turn attribute to greater resistance to Cl⁻ ingress.

On the other hand, Cong et al. [6] found that SF is more reactive than GGBS due to its high fineness by nature, and has a beneficial effect in improving the microstructure of hydrated cement paste in the vicinity of coarse aggregates. Persson [7] observed that the filler effect of SF reduces the volume of macropores, densifies the cement matrix, and thus limits the ingress of Cl⁻ in concrete considerably.

The investigations reviewed above and most research work reported in the literature on Cl⁻ ingress in blended cement concretes have been performed in temperate climates (average ambient temperature, $T^* = 20°C$).

However, there are relatively fewer studies on these concretes in tropical (30°C) and hot (40°C) climates. Hence, the objectives of this study are to investigate the influence of water-binder ratio (w/b) which is equivalent to water-cement ratio for 100% PC concretes, replacement percentage of PC by GGBS or SF and fineness of GGBS on the resistance to Cl⁻ ingress and microstructure (in relation to the pore size distribution and strength development); and to provide guidelines on the use of GGBS and SF to enhance the performance of marine structures, taking into account the effect of ambient temperature in temperate, tropical and hot countries.

EXPERIMENTAL DETAILS

Materials

The cementitious materials used are Portland cement (PC), ground granulated blast-furnace slag (GGBS) and silica fume (SF), whereas the aggregates adopted are river sand and crushed granite. The properties of these materials, as well as the mix proportions of PC concretes and pastes investigated in this study, have been reported by Wong [1].

Chloride Diffusion Test

Concrete prisms (150x150x400 mm) are water-cured at 20±1°C, 30±2°C and 40±0.5°C for 7 days, after which they are immersed in sodium chloride (NaCl) solution with Cl⁻ concentration equivalent to that of seawater (19380 ppm Cl⁻) [8] at the respective temperatures. After a scheduled immersion period, the prisms are split in the lateral direction. The silver nitrate (AgNO₃) spray method [9,10] is performed to visually identify the depth of Cl⁻ penetration x_S, which is then used as a measurement of Cl⁻ ingress.

In this study, all the prisms are fully submerged throughout the test. Hence, the following equation based on Fick's 2nd Law [11], which is commonly used to predict the ingress of Cl⁻ under pure diffusion, is adopted:

$$x_S = 4\sqrt{D_c t^*} + a' \tag{1}$$

where, x_S =   depth of Cl⁻ penetration from the AgNO₃ spray method (m)
           D_c =   apparent Cl⁻ diffusion coefficient (m²/s)
           t^* =   exposure period (s)
           a' =   constant (m)

The total Cl⁻ concentration (c_T) profile is determined by drilling the concrete specimens incrementally along the intrusion depth (at intervals of 0-5 mm, 5-10 mm, 10-15 mm and so forth) and collecting the powder samples. The Volhard titration method in accordance with BS1881:Part 124:1988 [12], together with the adjustment by Lee et al. [13] and Wee et al. [14], are then carried out to measure the value of c_T at each depth interval.

Mercury Intrusion Porosimetry (MIP)

The pore size distribution, i.e. the cumulative pore volume vs pore radius curve, of PC pastes is measured by means of a mercury intrusion porosimeter (model Porosimeter 4000 by Carlo Erba Instruments). This technique is based on the behaviour of non-wetting liquids, such as mercury, which is forced into the pore system of hardened cement matrix materials by means of an external pressure. The pressure exerted is inversely proportional to the pore radius, from which the pore volume is obtained.

In the present study, a maximum mercury intrusion pressure of 2 x 10⁸ N/m² is applied. The contact angle adopted for mercury is 141.3°. The samples are treated with acetone for a day, oven-dried at 105±1°C to a constant weight, and then kept in a vacuum desiccator for another day before testing.

Cube Test (Compressive Strength Test)

The concrete specimens (100 mm cubes) are cast and moist-cured for a day at 30°C, and then divided into three lots to be water-cured at 20±1°C, 30±2°C and 40±0.5°C. At ages of 3, 7, 28, 56 (for 40°C only) and 91 days, the cubes are removed from the curing tank for compressive strength tests in accordance with SS78:Part A16:1987 [15] (BS1881:Part 116:Part 1983 [16]). Three cubes cured at a certain age are tested for each concrete mix.

RESULTS AND DISCUSSION

Chloride Ingress

Effect of temperature of immersion solution on chloride ingress

Figure 1 reveals that the depths of Cl⁻ penetration (x_S) of concretes exposed to NaCl solution (19380 ppm Cl⁻) for 78 weeks at 40°C are the highest, followed by those at 30 and 20°C. It is also observed that a higher temperature leads to an increase in D_c (Figure 2) calculated using Equation (1). These results indicate that the rates of Cl⁻ ingress in hot climates are the highest, followed by those in tropical and temperate climates.

N30, N40, N50, N60   =   100% PC concretes with w/b of 0.3, 0.4, 0.5 and 0.6 respectively;
SF5, SF10   =   concretes with 5 and 10% replacement of PC by SF respectively;
B455, B655, B855   =   concretes with 55% replacement of PC by GGBS of 4500, 6000 and 8000 cm²/g fineness respectively.

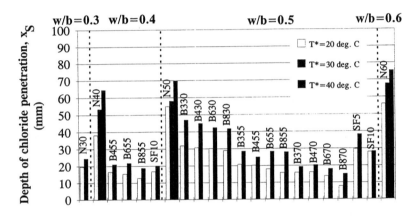

Figure 1 Depths of chloride penetration of PC concretes

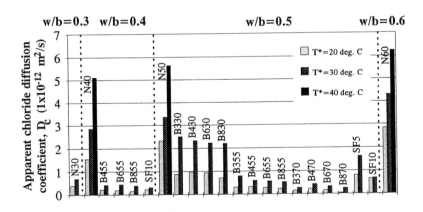

Figure 2 Apparent chloride diffusion coefficients of PC concretes

Effect of water-binder ratio and mineral admixtures on chloride ingress

It is found that improvement in the resistance to Cl⁻ ingress can be achieved by choosing a mix with a lower w/b (Figures 1 and 2) and/or using a higher replacement percentage of PC by GGBS or SF (Figures 1 to 4). However, an increase in the fineness of GGBS has less influence on this improvement (Figures 1 to 4).

Initiation period for chloride-induced steel corrosion in concrete

In the present work, the initiation period t_i [17] and a threshold total Cl⁻ concentration c_T of 0.40 wt% of cement [18] are adopted to assess the progress of activation front for chloride-induced corrosion of steel reinforcement in RC structures. The concrete cover (x_C) for a certain t_i is determined by intersecting a c_T profile with the threshold level of $c_T = 0.40$ wt% of cement, as illustrated in Figure 3. The relationship between t_i and x_C is then fitted to:

$$t_i = a(x_C)^b \qquad (2)$$

which can also be written as:

$$\log t_i = \log a + b \log x_C \qquad (3)$$

where, t_i = initiation period (weeks)
  x_C = concrete cover (mm)
  a, b = constants

The values of a and b of PC concretes with and without GGBS or SF, ranging from 0.0257 to 0.2323 and 2.0281 to 2.1458 respectively, are calculated by linear regression based on the log t_i vs log x_C data (Figure 4) obtained in this study. The correlation coefficients range from 0.9904 to 0.9999, indicating that the data points obey the power law relationship in eq. (2). Figure 4 reveals that for the same concrete cover (x_C), the values of t_i for PC concretes with GGBS or SF are larger than that of 100% PC concrete. This is likely to be caused by the pozzolanic reaction that takes place in concretes with GGBS or SF, resulting in a denser microstructure, a slower rate of Cl⁻ ingress, and thus a longer time to initiate steel corrosion in concrete. It is also observed that a higher replacement percentage of PC by GGBS leads to an increase in t_i as a result of a slower rate of Cl⁻ ingress. However, a higher fineness of GGBS is found to have less influence on t_i.

Microstructure of Concrete in Relation to Chloride Ingress

Effect of temperature of immersion solution on pore size distribution

The pore size distributions of PC pastes, having water-cured for 7 days and soaked in NaCl solution (19380 ppm Cl⁻) for 13 weeks, are obtained using MIP as shown in Figures 5(a) to (c). It is observed that 100% PC paste (Figure 5(a)) with a w/b of 0.5 and PC pastes (w/b = 0.5) containing GGBS (Figures 5(b) and 5(c)) at 30°C have smaller capillary pore volumes than at 20°C in the range of 100 to 100000 Å. This indicates that a higher temperature results in a denser microstructure.

N40   =   100% PC concrete with a w/b of 0.4;
SF10 =   concrete (w/b = 0.4) with 10% replacement of PC by SF;
B465 =   concrete (w/b = 0.4) with 65% replacement of PC by GGBS of 4500 cm²/g.

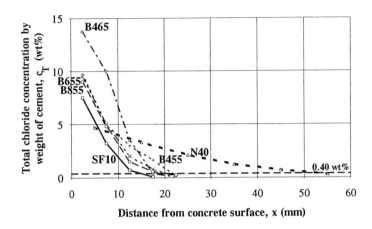

Figure 3   Total chloride concentration profiles of PC concretes (w/b = 0.4) after an immersion period of 78 weeks (1.5 years) at 30°C

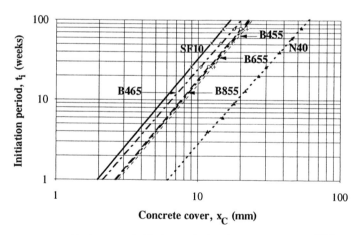

Figure 4   Relationship between initiation period and concrete cover of PC concretes (w/b = 0.4) at 30°C

Effect of water-binder ratio and mineral admixtures on pore size distribution

The cumulative pore volumes of 100% PC pastes cured at 20 and 30°C in Figure 5(a) are found to be smaller at a lower w/b, suggesting that at these temperatures, a lower w/b leads to a denser microstructure. It is also observed in Figures 5(b) and 5(c) that the cumulative pore volume decreases with increasing replacement percentage and fineness of GGBS.

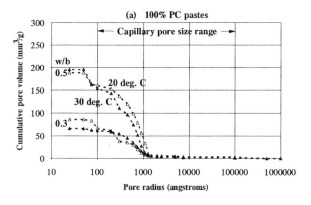

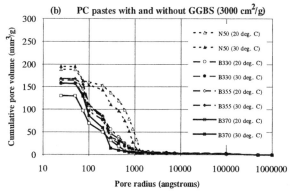

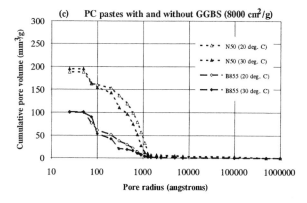

Figure 5 Pore size distributions of PC pastes, pre-cured in water for 7 days and immersed in
NaCl solution (19380 Cl⁻) for 13 weeks (0.25 year)

Effect of curing temperature on strength development

Figure 6(a) shows that the early compressive strength up to 28 days for 100% PC concretes cured at 40°C is the highest, followed by those cured at 30 and 20°C. On the other hand, the compressive strengths of concretes cured at 20, 30 and 40°C for 91 days are almost similar. Hence, as compared to the same concretes cured in tropical climates (30°C), concretes cured in hot climates (40°C) can be expected to have a higher early strength (28-day compressive strength), but an almost similar ultimate strength (91-day compressive strength). The same trend is also observed for concretes cured in tropical climates (30°C) with respect to those cured in temperate climates (20°C).

In Figure 6(a), the compressive strength curves of 100% PC concretes at 20°C are found to level off around 28 days. However, the curves of these materials at 30 and 40°C level off from 7 days, except for a lower w/b of 0.3 at 30°C of which the curve continues to develop after 7 days. This may be due to the curing temperature having little influence on PC concretes of a low w/b such as 0.3 and below, since the rate of hydration is more likely to be dependent on the availability of pore water to sustain hydration. For PC concretes (w/b = 0.5) with SF in Figure 6(b) or GGBS in Figure 6(c) at 20°C, the curves continue to increase up to 91 days but at 30°C, the curves level off after 28 days. This suggests that the hydration and pozzolanic reactions at a higher temperature (30°C) are relatively faster and near to completion at an earlier curing age than those at a lower temperature (20°C).

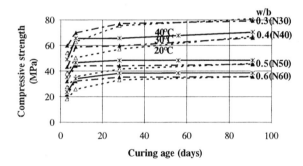

Figure 6(a) Strength development of 100% PC concretes

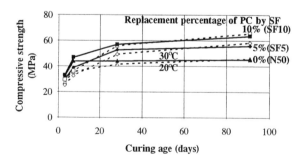

Figure 6(b) Strength development of PC concretes (w/b = 0.5) with and without SF

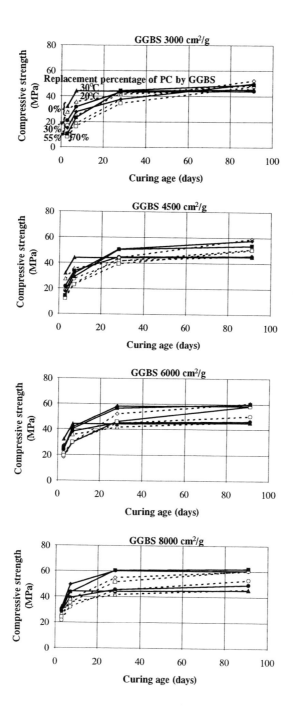

Figure 6(c) Strength development of PC concretes (w/b = 0.5) with and without GGBS

CONCLUSIONS

1. At a higher curing temperature, the early compressive strengths of concretes are greater, whereas the ultimate strengths are almost similar. The compressive strength curves also level off after a shorter curing age, suggesting that the hydration reaction is faster and approaches completion at an earlier curing period.

2. The improvement in the performance of concrete is generally achieved by adopting a lower water-binder ratio (0.3 to 0.6) and a higher replacement percentage of PC by GGBS (30 to 70%) or SF (5 to 10%). The use of GGBS with higher fineness (3000 to 8000 cm^2/g) improves the compressive strength, but has no significant effect on the resistance to chloride ingress. A decrease in water-binder ratio also leads to greater early and ultimate compressive strengths, as well as a denser microstructure.

3. The rate of chloride ingress is dependent on ambient temperature. This implies that there may be a need to consider the influence of climatic conditions in the design of marine structures, especially in tropical and hot countries.

ACKNOWLEDGEMENTS

The authors gratefully acknowledge the financial support (project GR6171) and supply of cementitious materials by Pan Malaysia Cement Works (Singapore) Pte Ltd. Special thanks are also extended to the technical staff of Concrete Technology Laboratory, Department of Civil Engineering, The National University of Singapore, for their assistance.

REFERENCES

1. WONG, S F. Ingress of Chloride in Concrete, PhD thesis, The National University of Singapore, 1998.

2. NAGATAKI, S AND TAKADA, M. Effects of interface reaction between blast furnace slag and cement paste on the physical properties of concrete. Proc. of the 7th International Congress on the Chemistry of Cement, Paris, 1980, pp III-7/90 - 7/94.

3. DHIR, R K, JONES, M R, AHMED, H E H AND SENEVIRATNE, A M G. Rapid estimation of chloride diffusion coefficient in concrete. Mag. Concr. Res., Vol. 42, No. 152, 1990, pp 177-185.

4. BYFORS, K. Influence of silica fume and fly ash on chloride diffusion and pH values in cement paste. Cem. Concr. Res., Vol. 17, No. 1, 1987, pp 115-130.

5. COLLEPARDI, M, MARCIALIS, A AND TURRIZIANI, R. Penetration of chloride ions into cement pastes and concretes. J. Amer. Ceram. Soc., Vol. 55, No. 10, 1972, pp 534-535.

6. CONG, X, GONG, S, DARWIN, D AND McCABE, S L. Role of silica fume in compressive strength of cement paste, mortar and concrete. ACI Mater. J., Vol. 89, No. 4, 1992, pp 375-387.

7.   PERSSON, B. Seven-year study on the effect of silica fume in concrete. Advn. Cem. Bas. Mat., Vol. 7, Nos. 3/4, 1998, pp 139-155.

8.   ASTM DESIGNATION : D1141-90. Standard specification for substitute ocean water. Annual Book of ASTM Standards - Water and Environmental Technology, Vol. 11.02, 1992, pp 439-440.

9.   WEE, T H. Presence of Chlorides in Hardened Cement Matrix Materials, DEng thesis, Tokyo Institute of Technology, 1990.

10.   ITALIAN STANDARD. Determination of the Chloride Ion Penetration, UNI 79-28, Rome, 1978.

11.   CRANK, J. The Mathematics of Diffusion, 2nd edn, Clarendon, Oxford, 1975.

12.   BRITISH STANDARDS INSTITUTION. Method for Analysing Hardened Concrete (BS1881:Part 124:1988), London, 1988.

13.   LEE, S L, WONG, S F, SWADDIWUDHIPONG, S, WEE, T H AND LOO, Y H. Accelerated test of ingress of chloride ions in concrete under pressure and concentration gradients. Mag. Concr. Res., Vol. 48, No. 174, 1996, pp 15-25.

14.   WEE, T H, WONG, S F, SWADDIWUDHIPONG, S AND LEE, S L. A prediction method for long-term chloride concentration profiles in hardened cement matrix materials. ACI Mater. J., Vol. 94, No. 6, 1997, pp 565-576.

15.   SINGAPORE STANDARDS. Methods for Determination of Compressive Strength (SS78:Part A16:1987), Singapore, 1987.

16.   BRITISH STANDARDS INSTITUTION. Method for Determination of Compressive Strength of Concrete Cubes (BS1881:Part 116:1983), London, 1983.

17.   TUUTTI, K. Service life of structures with regard to corrosion of embedded steel. Performance of Concrete in Marine Environment, SP65, Ed. P K Mehta, American Concrete Institute, Detroit, 1980, pp 223-236.

18.   BRITISH STANDARDS INSTITUTION. Structural Use of Concrete - Code of Practice for Design and Construction (BS8110:Part 1:1985), London, 1985.

CONCRETE SURFACES PRODUCED WITH CPF: DURABILITY AND EFFECT OF LINER REUSE

R E Beddoe

Technical University of Munich

Germany

ABSTRACT. Tests have been conducted to assess the durability of concrete surfaces produced using controlled permeability formwork (CPF) liners. Specimens prepared using Portland cement concrete with consistencies between plastic and flowable were exposed to outdoor conditions for 30 months. The carbonation depth of specimens prepared with liners was only 0.5mm, compared to depths up to 7mm for concrete cast against conventional formwork. The water and air permeability of the near-surface concrete produced with liners corresponded, irrespective of initial w/c ratio, to a concrete with a w/c ratio of approximately 0.40. The enhanced resistance against water and air penetration, as well as freezing and thawing in the presence of deicing salt, did not deteriorate during weathering. The performance of multiple-use liners was tested with Portland blastfurnace cement concretes of regular consistency. Smooth surfaces with few blowholes were produced with liners used three times. Sufficient drainage water was removed with used liners to increase surface strength by a DIN 1045 strength class, improve permeation characteristics, and reduce the depth of carbonation after 26 months by 50% to 3 mm.

Keywords: Controlled permeability formwork, Near-surface concrete, Durability, Carbonation, Water permeability, Air permeability, Freeze-thaw/deicing salt resistance

Dr R E Beddoe is in charge of Materials Physics at the Institute of Building Materials, Technical University of Munich where he lectures on the physics of engineering materials. His research specialises in the microstructure of binder and concrete with regard to materials properties, in particular, durability and transport. The effect of formwork liners on concrete quality is a main research topic. Other fields of interest include the hygral properties of hardened cement paste, small-angle X-ray scattering, rheology and non-destructive testing methods.

INTRODUCTION

Formwork liners are used to increase the durability of concrete structural members and produce concrete surfaces with few blowholes. This is achieved by the removal of excess air and water from the fresh concrete next to the liner. The commercial formwork liners presently available remove water from fresh concrete either by absorption or, in controlled permeability formwork (CPF), by acting as a filter membrane through which water is removed by the hydrostatic pressure of the fresh concrete. In general, CPF systems can be classified in three groups [1]: Type I: two-layer cloth system comprising filter and drainage layers, Type II: single layer filter system, Type III: two-layer system consisting of a filter membrane fixed to a drainage grid backing.

In Germany, the Type II liner Zemdrain produced by DuPont and its later Type III version Zemdrain MD, designed for multiple use, are widely employed to improve the durability of exposed areas of bridges, dams, dykes etc. and to produce smooth abrasion resistant surfaces of, for example, water storage tanks [e.g. 2]. According to earlier investigations [3] with the Type II liner, water drainage reduces the w/c ratio of a surface zone 2 cm in thickness to approximately 0.40. It was found that the lower w/c ratio, and therefore denser and less porous structure, of the surface zone is not just caused by water removal, but also by cement accumulation due to the hydrodynamic transport of cement particles in the fresh concrete towards the liner. This explains the increase in surface strength and freeze-thaw/deicing salt resistance as well as the reduction in carbonation, water absorption, chloride ingress and water and air permeability of concrete produced with this liner [e.g. 3-7]. A summary of indicative improvement levels is given in [1] for the surface properties of concrete made with CPF relative to conventional formwork.

Since the improved quality of the surface concrete should be maintained during the service life of structural components, the present paper is focused on the long term resistance of surface concrete produced with CPF against carbonation, water and air permeation and frost-deicing salt attack. Owing to its financial significance, the possibility of using a liner more than once without forfeiting concrete quality is a topic of growing importance. Results are presented on the effect of Type III CPF reuse on surface strength, permeability and carbonation.

EXPERIMENTAL DETAILS

Preparation of Specimens

A German Portland cement CEM I 32.5 R and a Portland blastfurnace cement CEM III/A 32.5 were used with local Munich aggregate of the grading A16/B16. The compositions of the mixes are listed in Table 1. The amount of cement paste in the fresh concrete was the same for Mixes 1 to 4, lower for 5 and 6, respectively, and the same for 7 and 8. The consistency of the fresh concrete was determined by measuring its spread according to the flow test DIN 1048 Part 1. A superplasticizer (1% with respect to cement content) was necessary to obtain regular consistency (spread between 42 and 48 cm) of Mix 7.

Table 1 Concrete mixes

CPF	MIX	CEMENT TYPE	W/C RATIO	CEMENT (kg/m³)	SPREAD (cm)
Type II[a]	1	CEM I 32.5 R	0.50	376	41.0
"	2	"	0.55	355	48.5
"	3	"	0.60	335	52.5
"	4	"	0.65	318	59.0
"	5	"	0.55	342	46.0
"	6	"	0.55	329	41.0
Type III[b]	7	CEM III/A 32.5	0.45	402	40.0
"	8	"	0.55	355	43.0

[a] Zemdrain, [b] Zemdrain MD

Special moulds were constructed from non-absorptive polymer-coated plywood, Figure 1. A CPF liner was fixed to one of the vertical 40×40 cm² sides of the mould so that the drainage water could escape through the joint between the base and side of the mould. All the other joints were sealed with silicone. The Type II liner Zemdrain was used in the investigations with Portland cement concretes and the multiple-use Type III liner Zemdrain MD was taken for the blastfurnace cement concretes. A release agent was, as specified by the manufacturer, not necessary with these liners.

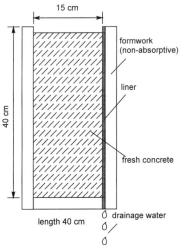

Figure 1 Mould used in the investigations

After pouring, the fresh concrete was compacted with an internal poker vibrator. The top layer of concrete was revibrated after 30 minutes. The specimens were demoulded after one day and stored at 20°C and a relative humidity of 65%. Each 40×40×15 cm³ block possessed a 40×40 cm² test surface cast against a liner and, on the opposite side, a reference surface produced without a liner.

TEST PROCEDURE

In order to determine the drainage capacity of the liners for each mix, the mixing water, which dripped for up to 4 hours out of the moulds after placing, was collected and its volume measured.

Type II Liner

At an age of two weeks the specimens were placed on the laboratory roof, with the test surfaces facing south and inclined at 45°, where they were subjected for 30 months to outdoor climatic conditions in Munich. Since the test surfaces were uppermost, and therefore more exposed to weather conditions than the reference surfaces on the opposite sides, a control specimen, made without a liner and using Mix 4, was included in the investigations.

After exposure, the surfaces were inspected for signs of erosion and the depths of carbonation determined according to the German guideline [8] by spraying a pH-sensitive phenolphthalein solution on the freshly fractured surfaces. The air permeability of the surface concrete was determined according to Schönlin [9]. A suction cap, 5 cm in diameter, was placed on the concrete surface, evacuated and the permeability calculated from the rate of pressure increase inside the cap. Specimens measuring $22 \times 22 \times 15$ cm^3 were cut from the centre of the blocks for the measurement of water permeation volumes described in DIN 1048 Part 5. A constant water pressure of 0.5 N/mm^2 was applied for three days to a circular area, 10 cm in diameter, on the concrete surface. The volume of water was recorded which penetrated the surface.

Reuse of Type III Liner

In these investigations, the liners were used three times without cleaning or removing them from the formwork board between uses. The surface strengths of the specimens were determined with the type N Schmidt rebound hammer (DIN 1048 Part 2). A total of 16 measurements were taken at points distributed around the centre of each test surface. Data were also recorded in the same manner for the reference surfaces on the opposite block sides. The water penetration volumes were measured as described above. At a later date, the water pressure was reapplied, the specimens split and the penetration depths measured. The halved specimens were then subjected to natural outdoor exposure for 12 months after which the depths of carbonation and air permeability were measured.

RESULTS

Drainage Water

Figure 2 shows the quantities of drainage water collected as a function of spread. The spread is divided into the standard consistency regions KP plastic, KR regular and KF flowable. Data are also shown for a number of additional mixes. The excess mixing water dripped out of the moulds until initial setting terminated the drainage process. It is evident that the amount of mixing water removed from the fresh concrete was determined primarily by consistency. The quantity of drainage water is a measure of the improvement of the near-surface concrete with respect to the interior. Very little drainage water left a mould containing the additional mix (a) indicating a w/c ratio close to 0.40 for the near-surface concrete which is correspondingly denser and less permeable than the interior, see [3]. Furthermore, the data imply that drainage performance and thus the near-surface concrete quality were enhanced by adding more superplasticizer to the mix.

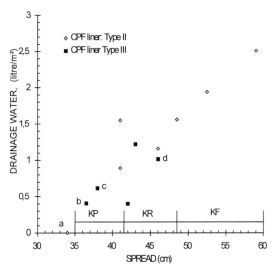

Figure 2   Drainage water as a function of spread measured in the flow test. Additional mixes: (a) Type II, w/c=0.40, 428 kg/m³ CEM I 32.5 R,(b, c, d) Type III, w/c=0.50, 380 kg/m³ CEM III/A 32.5 with 0.3, 0.5 and 0.8% superplasticizer.

In an earlier comparison of the effect of Type II and Type III liners on surface concrete quality [10], less drainage water was collected from Mix 8 with the Type III liner (1.22 litre/m²) than with the Type II liner (1.75 litre/m²). This difference was probably owing to the retention of drainage water in the grid backing of the Type III liner. Moreover, both liners proved to be equally effective in reducing carbonation. The water permeability and surface strength were even somewhat higher with the Type III liner. If, for a particular concrete composition, uncertainty exists regarding the improvement to near-surface concrete quality with CPF, a trial mould (Figure 1) should be constructed to measure the amount of drainage water. Further details on this recommendation are available in an earlier publication [5].

Effect of Liner Reuse: Each time the Type III liner was reused with Mix 8, the amount of water leaving the mould decreased slightly until, on the third application, the drainage water was 9% less than with a new liner, Table 2. No equivalent reduction could be determined for the stiffer Mix 7 owing to the small quantities of water collected (ca. 60 ml).

Visual Inspection of Surfaces

In contrast to the reference surfaces, the test surfaces produced by the Type II liner and multiple use of the Type III were very smooth and virtually free of blowholes.

Table 2 Collected drainage water for multiple use of Type III liner
with Portland blastfurnace cement concretes

NO. OF TIMES USED	MIX 7, W/C=0.45 [litre/m²]	MIX 8, W/C=0.55 [litre/m²]
1	0.40	1.22
2	0.35	1.19
3	0.46	1.11

On the first use of both liners, a lighter mostly void-free area, similar in hue to the reference surfaces, several centimetres in width was visible directly below the top edge of the test surfaces. The width of the lighter area increased several centimetres after each reuse of the Type III liner. However with the Schmidt hammer, no significant difference in surface strength was found between the lighter area and the remaining surface. It was possible to remove, to some extent, the lighter area with a steel wire brush.

The test surfaces of all the exposed specimens showed no signs of erosion and were virtually unchanged apart from some fine network cracks which were discovered on the surface of the Portland cement concrete specimens subjected to 30 months exposure. They appeared as a dark discoloration, about 0.1 mm deep with a mesh size around 15 mm, in the thin surface layer of fine mortar and were probably due to shrinkage. Although the majority of the reference surfaces showed no outward signs of deterioration, frost damage, in the form of cracks and crumbling, was evident around the edges of the reference surface of the block produced with Mix 4 (w/c=0.65).

Depth of Carbonation

After 30 months, the test surfaces of all the Portland cement concretes were carbonated, irrespective of concrete composition, to a depth of 0.5 mm whereas the depths of carbonation of the reference surfaces were between 2.5 and 7 mm, depending on w/c ratio. Thus the Type II liner reduced the carbonation by a factor of, on average, 10. The upper surface of the control specimen, made without a liner, was carbonated to a similar degree as the lower surface, i.e. 5 and 6.5 mm, respectively. Consequently, the difference in carbonation rate between the test and reference surfaces was not on account of their exposure conditions.

In accelerated tests [5], similar rates of carbonation were observed for conventionally cast specimens with w/c=0.4 and CPF cast specimens with higher w/c ratios. This indicates a near-surface w/c ratio in the region of 0.4 for the CPF concrete.

Effect of Liner Reuse: Table 3 shows the depths of carbonation for the Portland blastfurnace cement concrete specimens after multiple use of the Type III liner.

Both the test and reference surfaces of the specimens made with this cement exhibited higher rates of carbonation than the specimens made with ordinary Portland cement.

Table 3 Depths of carbonation after 26 months for multiple use of Type III liner with Portland blastfurnace cement concretes

NO. OF TIMES USED	MIX 7, W/C=0.45(mm)	MIX 8, W/C=0.55(mm)
1	3	4
2	3	5
3	3	4
Without liner	7	9

The depths of carbonation were not appreciably affected by liner reuse, and were on average roughly 50% of the value for conventional formwork. Although relatively small quantities of drainage water were removed from Mix 7, the rate of carbonation was nonetheless significantly reduced - even for the third liner use.

The carbonation depth of an additional specimen produced using Mix 7 and cast without a liner was 7 mm for both surface orientations.

Water Permeability

In three days, between 15 and 20 ml water penetrated each of the test surfaces of the ordinary Portland cement concrete specimens exposed for 30 months. This was far below the volumes of 50 to 150 ml, depending on w/c ratio, measured for the reference surfaces. Both surfaces of the exposed control specimen were penetrated by large quantities of water (>150 ml).

In supplementary tests, similar penetration volumes were obtained for concretes with a w/c ratio of 0.55 cast with the Type II liner and concretes with a w/c ratio of 0.4 cast in non-absorptive formwork.

Effect of Liner Reuse: Although the penetration volumes tended to increase each time the Type III liner was used (Table 4), the water volumes and depths are all well below those of the conventionally cast surfaces. Thus significant improvements in water permeability were also gained with reused liners.

Table 4 Volume and depth of water penetration at ages of 5 and 8 months, respectively, for multiple use of Type III liner with Portland blastfurnace cement concretes

NO. OF TIMES USED	MIX 7, W/C=0.45		MIX 8, W/C=0.55	
	Volume (ml)	Depth (mm)	Volume (ml)	Depth (mm)
1	10	10	10	20
2	15	10	40	25
3	20	15	50	25
Without liner	60	40	410	80

Air Permeability

In spite of variations over the surface of individual specimens, it was established that the permeability of the exposed specimens produced with the Type II liner was, with around 10^{-8} m^3/s, reduced with respect to the reference surfaces. Similar to water permeability, the air permeability of the reference surfaces tended to increase with w/c ratio and ranged between 5×10^{-8} and 2.0×10^{-7} m^3/s. The present data were also compared with earlier results taken for the same specimens at the beginning and after one year of natural exposure. As opposed to the reference surfaces, exposure did not significantly reduced the permeability of the CPF cast surfaces.

Effect of Liner Reuse: The air permeability of the exposed Portland blastfurnace cement specimens increased for each use of the Type III liner, Table 5. In a similar manner to water permeability, the air permeability of the test surfaces was, for all three applications, considerably reduced with respect to the reference surfaces. The permeability of the latter was very high and could only be roughly estimated because of the extremely rapid pressure equalisation in the suction cap.

Table 5 Air permeability at an age of 26 months for multiple use of Type III liner with Portland blastfurnace cement concretes

NO. OF TIMES USED	MIX 7, W/C=0.45 (m^3/s)	MIX 8, W/C=0.55 (m^3/s)
1	4×10^{-8}	8×10^{-8}
2	8×10^{-8}	1.5×10^{-7}
3	1.5×10^{-7}	2.1×10^{-7}
Without liner	ca. 10^{-6}	ca. 10^{-6}

The effect of reuse of the Type II liner on the permeation properties of ordinary Portland cement concrete, w/c=0.55, tested at an age of 28 days has been investigated at the Queen's University of Belfast [7]. The permeability increases observed after each use were, in this case, well below the present values. This difference was presumably caused by the higher degree of carbonation of the Portland blastfurnace cement concretes used in the present work.

Freeze-Thaw / Deicing Salt Resistance

The freeze-thaw resistance an exposed sample (Mix 2) in the presence of a 3% NaCl solution was tested according to the Austrian standard ÖNORM B3303. The scaling measured after 56 daily freeze-thaw cycles ($\pm 20°C$) was very light, 0.15 kg/m^2, and not above the scaling for a non-exposed specimen of the same composition, tested at an age of 28 days. Severe surface scaling, 8 kg/m^2 was exhibited by a specimen of the same composition, produced without a liner. Thus according to this test, the resistance of concrete made with Type II CPF against freezing and thawing in the presence of deicing salt was unaffected by natural exposure and the fine network cracks mentioned previously.

Furthermore, other workers found that microcracks in the surface of a power station quay wall cast with the same liner were not detrimental to the freeze-thaw deicing salt resistance as measured with cores [11].

Surface Strength

Effect of Liner Reuse: The effect of multiple liner use on the surface strength of Portland blastfurnace cement concrete is shown in Table 6. The average rebound distance measured with the Schmidt hammer is given in terms of scale divisions (sd). The standard deviation of the recorded values was 2.6 sd. The table also includes the corresponding strength classes according to DIN 1045 which were determined from the minimum average rebound distances specified by DIN 1048 Part 2.

Table 6   Average rebound distances at an age of 4 months for multiple use of Type III liner with Portland blastfurnace cement concretes

NO. OF TIMES USED	MIX 7, W/C=0.45		MIX 8, W/C=0.55	
	(sd)	Strength Class	(sd)	Strength Class
1	45.5	B35	46.3	B35
2	43.0	B35	40.9	B25
3	42.6	B35	40.0	B25
Without liner	37.8	B25	35.2	B15

It is apparent that the strength of the near-surface concrete decreased with each liner reuse whilst remaining above the strength of the corresponding reference surfaces. In all cases the surface strength was raised by at least a strength class. Similar improvements in tensile surface strength have also been obtained in pull-off tests on Portland cement concrete surfaces prepared with the Type II liner [7]. Furthermore, surface strength investigations [6] with a finer Portland blastfurnace cement of higher slag content (CEM III/B 42.5 - NW/HS/NA) showed similar improvements on first use of the Type II liner. In this case, the 28 day surface strength was improved by two classes from B35 to B55.

CONCLUSIONS

Smooth surfaces with few blowholes were produced with a Type II CPF liner for ordinary Portland cement concretes with consistencies between plastic and flowable according to DIN 1048 Part 1. The liner reduced the depth of carbonation after 30 months natural exposure by a factor of 10 to 0.5 mm. The rate of carbonation and the water and air permeability of surface concrete produced with the liner were similar for the range of Portland cement concrete compositions investigated. This was because water drainage, which was primarily governed by consistency, reduced the w/c ratio of the near-surface concrete to around 0.40, irrespective of its initial value, producing a dense surface zone.

Natural exposure did not reduce the enhanced resistance of the surface concrete produced by the liner against water and air penetration as well as frost and deicing salt attack. Very fine network cracks, which appeared as a dark discoloration in the thin surface layer of fine mortar during exposure, were not detrimental to durability.

Smooth blowhole free surfaces were produced for three applications of a multiple-use Type III CPF system to Portland blastfurnace cement concretes having regular consistency and w/c ratios of 0.45 and 0.55. The quantity of drainage water collected tended to decrease each time the liner was reused. Correspondingly, the water and air permeability increased and surface strength decreased with reuse. However in all cases, sufficient water was removed from both concretes to significantly lower water and air permeability, relative to surfaces produced with conventional non-absorptive formwork, and raise the surface strength by a strength class according to DIN 1045. The depth of carbonation of the Portland blastfurnace cement concrete was reduced by about 50% to between 3 and 4 mm after 26 months.

It should be pointed out that almost ten years have passed since concrete was first cast in Germany with the Type II liner. Thus a valuable source of in situ data for long term carbonation depths exists which ought to be exploited. Finally, formwork liners should not be used to rectify the adverse effects of an inappropriate concrete composition on durability, but to enhance the quality of surface concrete under severe conditions or for special requirements.

ACKNOWLEDGEMENTS

The author expresses his gratitude to the Deutsche Forschungsgemeinschaft for the financial support of part of the investigations. Special thanks is due to DuPont de Namours (Luxembourg) S.A. for funding and granting permission to publish results.

REFERENCES

1. PRICE, W.F., Recent developments in the use of controlled permeability formwork. Concrete, March 1998, pp 8-10

2. ZENITI, M., Preserving a Better Brew. Concrete Engineering International, March 1998, pp 43-45

3. BEDDOE, R.E., The Effect of Formwork Linings on Surface Concrete, Deutscher Ausschuß für Stahlbeton, 28. Research Colloquium, Technical University of Munich, October 1993, pp 109-113

4. KARL, J.-H. and SOLACOLU, C., Verbesserung der Betonrandzone. Beton, Vol. 43, No. 5, 1993, pp 222-225

5. BEDDOE, R.E. and SPRINGENSCHMID, R., The Influence of Formwork Inserts on the Durability of Concrete. Concrete Precasting Plant and Technology, Vol. 61, No. 2, 1995, pp 80-88

6.   BILGERI, P., Hoher Frost-Tausalz-Widerstand von Beton bei Verwendung von hüttensandreichem Zement und Zemdrain. Wülfrather Betonprüfstelle, Essen-Kupferdreh, Untersuchungsbericht Nr. 2, 1998

7.   LONG, A.E., BASHEER, P.A.M., BRADY, P. and Mc.CAULEY, A., Comparative Study of Three Types of Controlled Permeability Formwork Liner. International Congress, University of Dundee, June 1996

8.   DEUTSCHER AUSSCHUSS FÜR STAHLBETON. Prüfung von Beton. Empfehlungen und Hinweise als Ergänzung zu DIN 1048. DAfStb,Vol. 422, 1991

9.   SCHÖNLIN, K., Permeabilität als Kennwert der Dauerhaftigkeit von Beton. Schriftenreihe des Instituts für Massivbau und Baustofftechnologie, Karlsruhe Universität, Heft 8, 1989

10. SPRINGENSCHMID, R. and BEDDOE, R.E., Comparison of Concrete Surfaces made with the Formwork Liners "Zemdrain" and "Zemdrain MD". Report No. 3360-03-96, Institute of Building Materials, Technical University of Munich, 1996

11. WIES, S. and NEUBERT, B. Einfluß der Zemdrain - Schalungsbahn auf den Frost-Tausalz-Widerstand eines Betons. Forschungs- und Materialprüfungsanstalt, Untersuchungsbericht 12-21603/Wi/Nbt/H, 1995

INDEX OF AUTHORS

SUBJECT INDEX

This index has been compiled from the keywords assigned to the papers, edited and extended as appropriate. The page references are to the first page of the relevant paper.

288